Yulia Sandamirskaya

Sequence Generation in Dynamic Field Theory

Yulia Sandamirskaya

Sequence Generation in Dynamic Field Theory

Südwestdeutscher Verlag für Hochschulschriften

Impressum/Imprint (nur für Deutschland/only for Germany)
Bibliografische Information der Deutschen Nationalbibliothek: Die Deutsche Nationalbibliothek verzeichnet diese Publikation in der Deutschen Nationalbibliografie; detaillierte bibliografische Daten sind im Internet über http://dnb.d-nb.de abrufbar.
Alle in diesem Buch genannten Marken und Produktnamen unterliegen warenzeichen-, marken- oder patentrechtlichem Schutz bzw. sind Warenzeichen oder eingetragene Warenzeichen der jeweiligen Inhaber. Die Wiedergabe von Marken, Produktnamen, Gebrauchsnamen, Handelsnamen, Warenbezeichnungen u.s.w. in diesem Werk berechtigt auch ohne besondere Kennzeichnung nicht zu der Annahme, dass solche Namen im Sinne der Warenzeichen- und Markenschutzgesetzgebung als frei zu betrachten wären und daher von jedermann benutzt werden dürften.

Verlag: Südwestdeutscher Verlag für Hochschulschriften GmbH & Co. KG
Dudweiler Landstr. 99, 66123 Saarbrücken, Deutschland
Telefon +49 681 37 20 271-1, Telefax +49 681 37 20 271-0
Email: info@svh-verlag.de

Zugl.: Bochum, Ruhr-Uni, Diss., 2010

Herstellung in Deutschland:
Schaltungsdienst Lange o.H.G., Berlin
Books on Demand GmbH, Norderstedt
Reha GmbH, Saarbrücken
Amazon Distribution GmbH, Leipzig
ISBN: 978-3-8381-2472-8

Imprint (only for USA, GB)
Bibliographic information published by the Deutsche Nationalbibliothek: The Deutsche Nationalbibliothek lists this publication in the Deutsche Nationalbibliografie; detailed bibliographic data are available in the Internet at http://dnb.d-nb.de.
Any brand names and product names mentioned in this book are subject to trademark, brand or patent protection and are trademarks or registered trademarks of their respective holders. The use of brand names, product names, common names, trade names, product descriptions etc. even without a particular marking in this works is in no way to be construed to mean that such names may be regarded as unrestricted in respect of trademark and brand protection legislation and could thus be used by anyone.

Publisher: Südwestdeutscher Verlag für Hochschulschriften GmbH & Co. KG
Dudweiler Landstr. 99, 66123 Saarbrücken, Germany
Phone +49 681 37 20 271-1, Fax +49 681 37 20 271-0
Email: info@svh-verlag.de

Printed in the U.S.A.
Printed in the U.K. by (see last page)
ISBN: 978-3-8381-2472-8

Copyright © 2011 by the author and Südwestdeutscher Verlag für Hochschulschriften GmbH & Co. KG and licensors
All rights reserved. Saarbrücken 2011

Contents

List of Figures   v

List of Tables   ix

1 Introduction   1
- 1.1 Preamble 1
- 1.2 Motivation for the thesis 2
  - 1.2.1 Motivation within dynamical systems approach to cognition . . . 2
  - 1.2.2 Motivation within the problem of serial order. 4
- 1.3 Main goals of the thesis 6
- 1.4 Publications of the results of the thesis 7
- 1.5 Structure of the thesis 7

2 Background and methods   9
- 2.1 State of the art in modeling sequence generation 9
  - 2.1.1 Behavioral signatures of serial order 10
  - 2.1.2 Neurophysiology of serial order 11
  - 2.1.3 Classification of models of serial order according to Henson . . . 12
  - 2.1.4 Connectionist models 13
  - 2.1.5 Neural dynamic models 14
  - 2.1.6 Sequence generation in robotics 21
  - 2.1.7 Open issues in modeling sequential action and the problem statement for the thesis 22
- 2.2 Dynamic Field Theory to cognition 22
  - 2.2.1 Dynamic equation of a neural field 22
  - 2.2.2 Dynamic regimes and instabilities of a dynamic neural field . . . 23

CONTENTS

    2.2.3 Dynamics of discrete neural nodes. 25

    2.2.4 Coupling between neural fields and nodes 26

    2.2.5 Problem of sequentiality in DFT to cognition 27

3 The DFT sequence generation architecture. A one-dimensional model. 29

3.1 The task setting 29

3.2 The DFT architecture for sequence generation 30

    3.2.1 Ordinal fields 30

    3.2.2 Sequence memory: preshape of the ordinal fields 32

    3.2.3 Output field 32

    3.2.4 Condition of satisfaction system 33

3.3 Mathematical description of the model 35

3.4 Robotic implementation of the DFT sequence generation architecture . 36

3.5 Results of robotic demonstrations 40

    3.5.1 Learning 40

    3.5.2 Sequence generation 40

3.6 Discussion 43

4 Application of the DFT sequence generation mechanism to model turn taking 45

4.1 Embodied communication and turn taking: motivation for this application 45

4.2 The DFT model of turn taking 46

4.3 Mathematical description of the turn taking model 49

    4.3.1 The sequence generation mechanism. 49

    4.3.2 The turn taking mechanism. 50

4.4 Results 51

4.5 Discussion of the turn taking model 53

5 Ordinal dynamics based on discrete activation variables 57

5.1 The extended DFT sequence generation architecture 59

5.2 Mathematical structure of the DFT sequence generation architecture with discrete ordinal nodes 61

    5.2.1 Ordinal nodes 61

    5.2.2 Action fields and synaptic projections 64

		5.2.3	Condition of satisfaction field	65
		5.2.4	The sequential transition as a cascade of instabilities	66
	5.3	Robotic implementation		69
		5.3.1	The action and condition of satisfaction systems	71
		5.3.2	The color-space field: Interface to sensors and motors	71
		5.3.3	Motor system	74
	5.4	Results: the model in action.		76
		5.4.1	Sequence learning	76
		5.4.2	Sequence production	76
		5.4.3	Timing of actions	81
		5.4.4	Flexibility of sequence generation: no problems with repetitions	81
		5.4.5	Noisy environments	81
	5.5	Discussion		83
6	**A multi-dimensional DFT sequence generation architecture**			**85**
	6.1	Motivation for the multidimensional DFT sequence generation architecture		85
	6.2	The architecture		86
		6.2.1	Ordinal nodes	87
		6.2.2	Action fields and synaptic projections	87
		6.2.3	Condition of satisfaction fields	87
	6.3	Mathematical description of the model		88
	6.4	Robotic implementation		89
		6.4.1	The robotic scenario	89
		6.4.2	Implementation	90
			6.4.2.1 Teacher interaction	90
			6.4.2.2 Perception and motors	91
		6.4.3	Results	92
	6.5	Discussion		93
7	**Application to control the cognitive dynamics within a DFT architecture for spatial language.**			**95**
	7.1	Spatial language: Motivation for the application.		95
	7.2	Spatial language architecture: robotic implementation.		97

CONTENTS

- 7.3 An extension of the spatial language architecture and the need for sequential organization of boosts 99
- 7.4 Mathematical description of the modified architecture with the ordinal component 103
  - 7.4.1 Color-space fields 103
  - 7.4.2 Target field 103
  - 7.4.3 Reference field 104
  - 7.4.4 Transformation field 104
  - 7.4.5 Object-centered field 105
  - 7.4.6 Spatial and color nodes 105
  - 7.4.7 Ordinal dynamics 106
- 7.5 Results: autonomous sequencing of boosts. 106
- 7.6 Discussion 111

8 Discussion 113
- 8.1 Addressing the goals set for the thesis 113
- 8.2 Overview of the results 115
- 8.3 Outlook 117

9 Materials & methods 121
- 9.1 Parameters and their values used in implementations the sequence generation model 122
- 9.2 Parameters and their values used in the implementation of the spatial language architecture with ordinal dynamics 123

References 125

List of Figures

2.1 A typical recall episode for a sequential pattern in an STDP neural network. 16
2.2 A decision-making dynamics (Rabinovich *et al.*, 2006) 18
2.3 Output of the sequence generating architecture of (Deco & Rolls, 2005) 20
2.4 Detection instability in a two-dimensional dynamic neural field. 24

3.1 Overview of the DFT architecture for generating action sequences . . . 31
3.2 Four stages of the sequence generation dynamics. 33
3.3 The robotic implementation of the DFT sequence generation architecture for the color-search task. 34
3.4 The DFT sequence generation architecture in a learning session of the color-search task. 35
3.5 Learning of a sequence of colors. The user shows color blocks to the robot in a given serial order. 37
3.6 Generation of a color-search action sequence in the robotic arena. . . . 37
3.7 Ridge-shape input from the output field of the DFT sequence generation architecture to a two-dimensional perceptual color-space field. 38
3.8 A learning session 41
3.9 Time-courses of activity of the sequence generating architecture during production of two exemplar sequences 44

4.1 The DFT turn taking architecture 48
4.2 Activity of the oscillators that model action systems of the actors 52
4.3 Histogram of the durations of silence intervals between turns observed experimentally and generated by the DFT turn taking model 53

LIST OF FIGURES

5.1 The DFT sequence generation models with representation of ordinal information within continuous neural fields and within a discrete set of ordinal nodes, an overview 58

5.2 Functional modules of the DFT sequence generation architecture. 59

5.3 Activation levels of the ordinal and memory nodes at three time slices during a transition in a sequence 62

5.4 Visualization of the ordinal dynamics at the same three time slices during a transition, as in the Figure 5.3 and the time course of ordinal nodes' activations 63

5.5 Projections from the ordinal nodes onto the dynamic field representing a neural parameter of action. 65

5.6 Detection instability in the condition of satisfaction (CoS) field. 67

5.7 A transition between two actions in a sequence 68

5.8 Robotic demontration of the sequence generation architecture on a Khepera robot equipped with a color camera 70

5.9 Perception of the robot 72

5.10 Sequence learning in the DFT architecture driven by the visual input presented by a user 77

5.11 Two snapshots of dynamics of the DFT sequence generating architecture during sequence production on a robot 79

5.12 Time-course of a robotic demonstration 80

5.13 Three demonstrations of sequence learning and production 82

6.1 The sequence generating architecture including action systems specific to the robotic implementation. 86

6.2 Snapshot of the dynamics of the model 88

6.3 The robotic scenario: a sequence "find yellow"-"lower gripper"-"close gripper"-"find green"-"lower gripper"-"open gripper" results in the robot transporting a yellow object and depositing it on a green object. 90

LIST OF FIGURES

6.4 Time-courses of the dynamics of ordinal nodes and the three action fields during learning and production of a multimodal sequence. Dark regions on the field plots correspond to a negative activation in neural fields, light stripes with a dark midline are traces of peaks of positive activation, light regions with a lighter gray around them are traces of localized subthreshold preshapes. 93

7.1 Overview of the DFT spatial language architecture. In the plots of *spatial semantic fields*, dark regions correspond to strongly negative activation of the neural fields, light gray regions – to less negative regions that receive spatial templates input mediated by the activity of spatial nodes and output of the reference field. In the plot of the *reference field*, lighter regions correspond to higher activation, and in the plots of *color-space fields*, dark regions correspond to high levels of activation in the middle plot (red originally), or to strongly negative levels of activation in the right plot (blue originally). The latter region stems from the negative input from the reference field. The architecture is explained in the main text. 97

7.2 Overview of the modified spatial language architecture. 100

7.3 Task "What color has the object to the right from the green one". Non-autonomous architecture 101

7.4 Task "Where is the red object relative to the green object?". Non-autonomous architecture 102

7.5 Task "Where is the red object relative to the green object?". Non-autonomous architecture. 107

7.6 Tasks "What color has the object to the right of the green object?" and "Where is the red object relative to the green object?". Boosts are preprogrammed and are introduced to the system at fixed time intervals. 109

7.7 Tasks "What color has the object to the right from the green object?" and "Where is the red object relative to the green object?". Boosts are introduced by the sequence generation system. 110

LIST OF FIGURES

List of Tables

9.1 Numerical values of parameters used in implementations of the sequence generation model. Units are displayed where applicable. 122

9.2 Numerical values of parameters used in the implementation of the spatial language architecture with ordinal dynamics. Units are displayed where applicable. 123

GLOSSARY

1
Introduction

1.1 Preamble

Every human activity – working at your desk, playing football, or cooking a meal – involves cognitive processes. These processes include detecting objects in the environment and selecting among them those relevant for the moment, making decisions to initiate actions, holding in memory locations where objects were stored, estimating objects' properties and poses to hold or hit them properly.

Embodied cognition studies these cognitive processes emphasizing their close link to the body and the environment in which cognition takes place. This body consists of a *nervous system* where processes such as memory, learning, perception, attention, recognition, and executive control reside, *sensory systems* that conduct environmental inputs to the neural 'hardware', and *motor systems* that impact on the environment under the control of neural processes. Taking embodiment seriously means to close the gap between the neural processes that occur at neural time scale, according to the laws of neural dynamics, and overt behavior that occurs on its own time scale in a physical environment, only partially known to the acting agent. To close this gap, stability of the cognitive states is needed. Real-world environments are dynamic and introduce varying inputs to the neural system that controls behavior. Moreover, the sensory systems introduce noise to the perceptual input. Competing processes might also disturb the current state. Stability of cognitive states enables coherent behavior in a physical environment.

Naturally, every human activity involves sequences of actions: "first grab the ball,

1. INTRODUCTION

then drop it, then hit it". Cognition itself consists of sequences of actions or thoughts. Action sequences also constitute a crucial aspect of artificial, robotic action. Stability is a problem when sequentiality of behavior is considered. In order to activate the next action, the previous action must be deactivated, thus lose its stability. Understanding how action sequences are generated under constraints of embodiment is a core element of understanding cognition. And this is what my thesis is about: in the framework of Dynamic Field Theory that offers the means to model cognitive processes under embodiment constraints, I introduce a model for learning and generating sequences of dynamically stable states that are coupled to sensory inputs, are subject to a dynamics that supports cognitive functions, and that are also coupled to motor systems. Several implementations of the model on a robotic hardware demonstrate how it can guide autonomous behavior in a physical environment. Implementations in cognitive architectures show how the model can be integrated into complex process models of different aspects of human cognition.

1.2 Motivation for the thesis

1.2.1 Motivation within dynamical systems approach to cognition

In the dynamical systems approach to cognition, the behavior of an agent is described by one or several characteristic variables. These variables can reflect perceptual states (color, orientation, brightness of an object in the visual stream), motor commands (desired heading direction, joint angle), or complex states of the environment (position relative to a landmark, shape) (Schöner, 2008). For instance, movement of a robotic vehicle moving on even ground can be described by its heading direction in certain coordinate frames, whereas a red object perceived by a person can be described by a sharp color distribution centered around the hue value of 10. Behavioral variables may be neurally represented by neurons broadly tuned to such dimensions. This level of description abstracts from individual biological neurons and their spiking behavior. Firing of a whole population of neurons corresponds to a given value of the behavioral variable.

There are two core assumptions in the dynamical systems approach to cognition, that constrain models developed in this framework. First, the characteristic behavioral variables evolve in time according to continuous-time dynamical system equation.

1.2 Motivation for the thesis

With this constraint we commit to the fact that natural systems evolve autonomously in continuous time. As natural as it is, this assumption is not used to constrain many cognitive and even neural models, and even less so robotic architectures (Anderson *et al.*, 2004; Arbib, 1998; Cooper & Shallice, 2006; Henson, 1998; Keele *et al.*, 2003; Mataric, 2002; O'Reily, 2006; Rosenbloom *et al.*, 1991; Toussaint & Goerick, 2010; Wolpert & Kawato, 1998). In the latter models, the system can evolve from one state to the next one in a discrete stepwise manner. For higher-level cognitive functions and behaviors, this abstraction seems adequate, as discrete entities can be segmented in the behavior – words in language, actions in a sequence, objects in a scene. How these entities emerge from the underlying continua of the physical world, perceived through sensory systems, is left unexplained in such approaches. On the neural side, because spiking events are typically asynchronous, the time at which activation states are obtained at the level of a large population varies continuously (Amari, 1977; Ermentrout, 1998; Wilson & Cowan, 1973). Continuous time naturally replaces event time in such an asynchronous picture. This dictates the need for a coherent theory to specify how the distinct, temporally discrete in time macroscopic behaviors (and percepts) emerge from continuous neural processes and unfold in real-time in physical environments.

Understanding how behavior unfolds in a physical environment that has its own dynamics requires the concept of stability, the second core assumption of the dynamical systems approach to cognition. Thus, we assume that the dynamics that controls the evolution of a characteristic behavioral variable has an attractor solution. Moreover, the overt behavior of an agent always corresponds to an attractor solution of this dynamics (Schöner, 2008; Schöner *et al.*, 1995; Spencer & Schöner, 2003; Thelen & Smith, 1994). The second assumption constrains the choice of the characteristic variables and the dynamics. In the dynamical systems approach to cognition, they must have the attractor property, as described above. As the agent behaves in a world, the particular attractor might become unstable and a new attractor might emerge. The transitions between two attractors are short and are not reflected in a meaningful overt behavior. This assumption is motivated by the fact that neural systems control the behavior of agents in the face of neural and sensory noise, unpredictable environmental influences, and inputs coming from interfering tasks and goals. In order to produce coherent behavior, as well as to form a percept that can have a coherent effect, dynamic stability is

1. INTRODUCTION

necessary. The stability property is confirmed by a broad ensemble of experimental evidence in motor control (Kelso, 1995; Kelso & Schöner, 1987), and perception (Schöner, 2008). Stability is pervasive in neural systems.

These two assumptions distinguish the dynamical systems approach to natural and artificial cognition from other approaches including connectionism (Spencer *et al.*, 2009) and information processing (Shiffrin & Schneider, 1977). The attractor property of the meaningful behavioral states in this approach poses a problem when a sequence of behaviors is considered. A stable state corresponding to a particular action must be destabilized and the new attractor must emerge in an instability. The signal for this transition has to be picked-up by real sensors, if action is executed in a partially observable environment. The dynamics holding representations of the relevant aspects of an environment must, therefore, be endowed with memory and learning mechanisms (Spencer & Schöner, 2003). Moreover, the serial order of the actions must be represented in a structure that enables learning and incorporation of additional constraints, such as rules of behavior organization. Dynamic Field Theory (DFT) is a framework within which these contraints can be realized (Schöner, 2008). A mechanism for destabilizing the attractor states in a DFT sequencing architecture is the main contribution of this work. In Section 2.1, I review alternative approaches to sequence generation and discuss why this problem is overlooked in these disembodied treatments.

1.2.2 Motivation within the problem of serial order.

Many cognitive processes depend on sequences of perceptual states, actions, or thoughts. Active behavior in an environment requires that sequences of motor actions are produced. Serial order among these states or actions is crucial for the success in many cognitive tasks. Think of solving a problem going through several logical steps, writing down a telephone number from memory, or performing any planned action, like a trip to Hawaii. Generating action sequences involves many different processes including creating working and long-term memories for the elements of the sequence, initiating any individual action while inhibiting other available actions, controlling and timing elementary actions, terminating a completed action and selecting the next action.

Modern theoretical work on sequence generation has addressed aspects of serial order, such as characteristic patterns of errors in serial order tasks, how response times depend on serial position, sequence length, and other factors across a range of tasks

1.2 Motivation for the thesis

including speech production (Dell *et al.*, 1997b; Hartley & Houghton, 1996), spelling (Glasspool & Houghton, 2005), action planning (Cooper & Shallice, 2000; Grossberg, 1978), immediate serial recall (Henson, 1998), and free recall (Farrell & Lewandowsky, 2002). In many cases, sequences have been studied and modeled in relatively disembodied form, in which item and order memory is probed through simple, invariant motor responses such as keyboarding or button pressing. Such sequences are typically highly constrained. For instance, playing a piece of piano music entails sequential key presses, the order and duration of which is predetermined by the score. What little variance remains is constrained by speed-accuracy trade-offs (Pfordresher *et al.*, 2007).

Embodied agents that are situated in real environments, on the contrary, must make variable movements oriented at physical objects to achieve the goal of each sequence element. In this case, five properties of embodied sequence generation must be considered when modeling these processes.

First, embodied sequence generation is *autonomous*, that is, decisions to bring a particular object into the foreground, select it as the target of an action, and initiate that action are all driven by intrinsic processes that are coupled to the embodied system's own sensory and motor surfaces.

Second, each action within an embodied sequence is *part of a motor and perceptual continuum*. The particular movements required in each step in a sequence vary each time depending on initial configuration of the agent and the environment. Any stage of the behavioral sequence thus really entails a continuum of possible percepts and movements out of which a specific instance is generated in the environmental context on each occasion. Although the perceptual objects and actions relevant to each stage of the sequence may be described as belonging to a particular category, their categorical representation alone is too poor to explain the observed behavioral flexibility when context varies.

Third, embodied sequence generation is *flexibly timed*. Even though each element in a sequence may take varying amounts of time depending on the current environmental situation, each action is brought to a conclusion before the next one is initiated. This is true also if an action suffers perturbations. Embodied sequence generation must be stable against variations of the time needed to terminate each action and must be able to use sensory information to determine that a subgoal of the sequence has been reached.

1. INTRODUCTION

Two additional constraints arise from the demand that an understanding of embodied sequence generation be based on neuronal principles. Neural processing occurs *continuously in time*, sampled by asynchronous neuronal activity and continuously coupled to sensory inputs. Neuronal processing does not consist of discrete computational steps from one action or perceptual state to the next. Because neuronal time is fundamentally continuous, the emergence of discrete transitions between different sequence elements is in need of explanation.

Finally, neuronal representations naturally capture the continua of possible actions and possible sensory states through *graded representations*. The fact that neurons are discrete entities does not lead to observable discreteness of behavior. This is the basis for the neuronal concept of population coding (Deadwyler & Hampson, 1995; Erickson, 1974; Georgopoulos, 1991). Conversely, there are behavioral signatures suggesting that the metric dimensions of perceptual and motor representations are encoded. For instance, drift of metric memory over delays and the dependence of performance on task metrics can be accounted for by using population coding ideas (Johnson *et al.*, 2008). How the categorical states that characterize the different stages of a reproducible sequence emerge from such underlying continua is thus in need of explanation.

Because current neuronal models of sequence generation (Botvinick & Plaut, 2006; Brown *et al.*, 2000; Burgess & Hitch, 1999; Deco & Rolls, 2005; Elman, 1990; Houghton, 1990; Page & Norris, 1998) address these constraints only partially, there is, to date, no comprehensive theoretical account for embodied and situated sequence generation that explains how behavioral sequences are acquired and produced in the real world. One of the goals of this thesis is to establish such an account.

1.3 Main goals of the thesis

The main goals of my thesis can be summarized as follows.

1. Introduce and analyze a model for sequence generation in Dynamic Field Theory (DFT) that solves in a principled fashion the fundamental problem of the stability vs. sequentiality trade-off.

2. Develop and implement an architecture based on the DFT sequence generation model that takes into account constraints of the embodiment on an autonomous robotic agent.

3. Demonstrate in robotic experiments the core properties of the DFT sequence generation model.

4. Apply the DFT sequence generation model to solve problems of serial order in cognitive tasks.

1.4 Publications of the results of the thesis

The first model of sequence generation within DFT has been published as a refereed contribution to a conference (Sandamirskaya & Schöner, 2008). An exemplary application of this architecture to model turn taking was published as a book chapter in the volume "Embodied Communication" (Sandamirskaya & Schöner, 2006). The second model, with discrete implementation of the ordinal system was recently published in the journal "Neural Networks" (Sandamirskaya & Schöner, 2010a), whereas the multimodal extension of this model was published as a refereed proceedings paper (Sandamirskaya & Schöner, 2010b) and as an abstract (Sandamirskaya & Schöner, 2009). The spatial language architecture was developed in collaboration with John Lipinski (Lipinski et al., 2009a,c,d; Sandamirskaya et al., 2010). The application of the sequence generation model to control the behavior organization in the spatial language architecture is introduced here for the first time.

1.5 Structure of the thesis

Chapter 2 of the thesis provides a review of the state of the art literature on sequence generation and an introduction to the Dynamic Field Theory (DFT). Chapter 3 presents a one-dimensional model to represent serial order and stabilize sequential transitions. In Chapter 4, an application of the one-dimensional architecture to model turn taking in a dialogue within the field of embodied communication is described. Chapter 5 describes a discrete version of ordinal dynamics that is the first step towards the multi-dimensional sequence generation model, presented in Chapter 6. Chapter 7 deals with an application of the discrete ordinal dynamics to control sequential activation of dynamical modules in a DFT spatial language architecture. I finalize the thesis with a Discussion and provide a short summary of results and an outlook in Chapter 8.

1. INTRODUCTION

2
Background and methods

2.1 State of the art in modeling sequence generation

Even the simplest activities in life involve generating well-ordered sequences of actions. Preparing a meal and setting the table, packing the briefcase and driving to the office, or even just getting up from your desk to refill your glass at the water cooler are examples. Playing music, writing, and gesturing all entail serially ordered sequences of actions. Sequences are also the basis for the highest forms of cognition such as generating and comprehending language. Naturally, how sequences are generated has been a central topic of psychology since Lashely's seminal insight that the serial order in a sequence is a separate and critical dimension of such behavior (Lashley, 1951).

How humans and cognitive robots generate sequences of actions is constrained by their embodiment and situatedness (Riegler, 2002). The physical properties of the body imply, for instance, that actions take characteristic amounts of time that may vary depending on circumstances. Physical environments may vary in time on their own characteristic time scale. Sensory and motor systems operate in continuous time and have graded state variables that are subject to spatio-temporal fluctuations. Sequence generation must be capable of integrating these constraints in order to produce stable behavior. The mechanisms of sequence generation, however, are often studied in isolation from the sensory-motor systems that drive learning of sequences of actions and their generation.

In the following I review some literature form the fields of psychology, neuroscience, and neural dynamics, highlighting some findings in these fields that have inspired my

2. BACKGROUND AND METHODS

work, but also emphasizing that in most cases the problem of embodiment of sequence generation has not yet been considered. In particular, neither of the five properties of an embodied system – autonomy, continuity of representations, stability, flexible timing, and gradedness, are addressed in the literature in the context of sequence generation.

2.1.1 Behavioral signatures of serial order

The large literature on theoretical models of serial order aims at understanding the structure of the underlying memory systems and use serial order errors and response times as signatures of sequence generation mechanisms. Thus, error patterns in language production (Dell *et al.*, 1997a,b), dependence of errors on speed of production (Pfordresher *et al.*, 2007), sequence length and structure (Deroost *et al.*, 2006; Erhorul & Eichenbaum, 2006) can be analyzed to gain insights in mechanisms of sequence generation.

These studies established, for instance, that the memory for items at the beginning and at the end of a sequence is more robust (primacy and recency effects) (Li *et al.*, 2000; Nairne, 1991), that repetitisms of particular items or similar segments facilitate errors, and that there are more errors in longer sequences. The most frequent errors in serial order tasks are exchanges of items in adjacent or close ordinal positions. Even more of these errors occur if the affected items are similar (Lee & Estes, 1977). If the items in a sequence are grouped, then exchanges may occur between items at the same ordinal position in different groups (Henson, 1998). Also there are positional errors between experimental trials: an erroneous item in one trial is more likely than chance to have occurred at the same position in the previous trial (Conrad, 1965; Glasspool *et al.*; Henson, 1998; Ryan, 1969). Other frequent serial order errors are omissions and insertions, alternation reversal, coarticulation, or anticipation, shifts and substitutions, and dissociation of the property of doubling from the item (Glasspool, 2005).

When modeling their experimental findings, the authors in this field are typically not concerned with the physical production of each action and the acquisition of relevant sensory information. They aim to explain empirical findings from experiments, where individual actions in a sequence are simple stereotypical motor acts, such as button pressing or typing. Different durations of actions are not taken into account and the links to sensory-motor systems are not specified. However, I developed the

2.1 State of the art in modeling sequence generation

DFT sequence generating model to be consistent with a theory that seems to address the behavioral data best, as discussed in Subsection 2.1.3.

2.1.2 Neurophysiology of serial order

Although much remains to be known about the neuronal basis of sequential behavior, neurons coding for ordinal position have been fond in a number of relevant neuronal structures (Tanji, 2001). Neurons in the supplementary motor area were found to become active for specific movements within a movement sequence but not when those movements were performed outside the sequence context (Shima & Tanji, 1998). In anterior cingulate cortex, some neurons showed activity selectively for a particular ordinal position but independently of which particular movement was being made at the upcoming, preceding, or subsequent stage of the sequence (Procyk et al., 2000). A similar pattern was found when investigating the difference in neural activations during "syntactic" and "natural" grooming sequences in rats (Aldridge & Berridge, 1998). A population of dorsolateral neurons in this study was active only during the particular sequential patterns of grooming movements and not when the same patterns occurred outside the grooming sequence. These populations thus appeared to encode the serial order and not the motor properties of constituent movements. Neural pools responsive to serial information were even found in motor cortex (Carpenter et al., 1999). The ordinal nodes of the DFT model are inspired by these neural signatures. Prefrontal regions play a role in the executive control of sequential action (Fujii & Graybiel, 2003). Prefrontal neural activity may sustain information about sequences between trials (Averbeck & Lee, 2007). Activity after a learning trial and before production of a sequence predicts which of a number of learned sequences is going to be initiated. Different patterns of neural activity were observed during the preparation of a movement sequence, before a particular ordinal position, and in specific time intervals between serial items (Tanji, 2001). Overall, neurophysiological findings provide convergent evidence for a positional account for serial order, in which ordinal information is coded in a separate representation that projects on the perceptual and motor representations of the items in a sequence.

The basal ganglia play a role in the temporal organization of actions. The basal ganglia neurons are active during both the learned motor or cognitive tasks and during learning of novel motor behavior (Hikosaka et al., 2000). The inhibitory connections

2. BACKGROUND AND METHODS

are established from the discrete regions of basal ganglia to separate cortical regions (Parent & Hazrati, 1995). As this structure is also considered to participate in reward anticipation mechanism, the function of the "condition of satisfaction" system presented below might resemble that of the basal ganglia structures.

Certainly, multiple mechanisms support production of sequentially ordered actions in natural and artificial systems. As I describe my model, I will keep track of the mapping between the model and known neurophysiological facts.

2.1.3 Classification of models of serial order according to Henson

A useful classification of the serial order models distinguishes between chaining, ordinal, and positional theories (Henson, 1998). Chaining theories postulate that serial order is stored in directional links between the successive states. Thus, chaining theories are attractive, as they use simple learning rules, are intuitive, and the mechanism of chaining may be relevant to behavioral organization, or play a role in habit formation and routine sequences (Botvinick & Plaut, 2006; Hikosaka *et al.*, 1999). A number of arguments can be advanced against chaining theories, however: limited flexibility is one of them, a problem with repeated items is another one (Cooper & Shallice, 2000; Dell *et al.*, 1997b; Henson, 1998). The latter can be overcome by introducing context-specific coding that considers several previous outputs in deciding for the next output (Jordan, 1997). Such context-specific coding implies, however, extensive learning of each sequence that must be remembered. Chaining theories are not considered as viable models to explain serial order errors, they are also not flexible enough to explain sequence generation in an acting agent.

Ordinal theories postulate that the order of an item in a sequence is represented in an activation gradient over a network that represents item information (Glasspool *et al.*, 2004; Grossberg, 1978; Page & Norris, 1998). The most active item is read out by a winner-take-all algorithm, inhibition of return provides for extinction of the activity of the currently selected item. Ordinal theories seem to contradict some experimental data on human (Henson, 1998) and animal (Eichenbaum, 2007) behavior. In the latter study, for instance, rats seem to store information about the order of odors directly, not in relation to the previous or the following item. The localized lateral inhibition that produces the gradient of activation in ordinal theories also causes problems similar to the ones of chaining theories.

2.1 State of the art in modeling sequence generation

To date, positional models are the most successful candidates in addressing behavioral data on serial order (Burgess & Hitch, 1999; Dell *et al.*, 1997b; Henson, 1998; Houghton & Hartley, 1995). Some error types – transpositions between groups of items and positional errors between trials – can only be accounted for within these theories. In positional theories, item information is associated with a location that encodes ordinal position. This makes it possible to both account for ordinal errors through disruptions within the positional representation of serial order as well as to account for similarity effects and the influence of the domain structure. However, most positional models of serial order do not provide a process model of sequential switching. Moreover, the ordinal code is often numerical, a neural counterpart of it is left unspecified. Timing of the items in a sequence is fixed by an external timer. The model presented in this thesis is technically a neural-dynamic instantiation of a positional theory for serial order – the ordinal information within the model is represented explicitly and is coupled to representation of the content at each ordinal position. This model, however, respects both the constraints of neural implementation and of acting out sequences in the physical world.

2.1.4 Connectionist models

The embodiment of sequence generation is, to a limited extent, addressed by architectures based on recurrent neural networks (Botvinick & Plaut, 2004; Dominey *et al.*, 1995; Elman, 1990). Here, the weights of a neural network are trained by exposing an agent to different behavioral sequences. The resulting network produces an appropriate sequence of outputs representing actions in a simulated environment. The environment provides input after each sequential action, which, combined with the internal state of the network, leads to the next output value. How that perceptual input is acquired from noisy environments is not addressed nor is the time course of action controlled and stabilized. Neural network models of this type have not been implemented in real-world agents and such implementation would require solving these problems. The recurrent neural network architectures do not model the process of acquiring an action sequence in a natural setting with a single or few demonstrations. The back-propagation algorithm, which these networks rely on, requires extensive learning under supervision. The models do address, however, empirical findings on routine sequences and may provide insights into action slips (Botvinick & Plaut, 2004).

2. BACKGROUND AND METHODS

Embodiment is only a secondary concern for the broader and more abstract neural dynamics models of sequence generation (Gnadt & Grossberg, 2007; Grossberg & Pearson, 2008) that are based on Grossberg's proposal of a neural mechanism for serial order (Grossberg, 1978). Here, sequences are stored as spatial gradients of neural activation. A volitionally-activated non-specific rehearsal wave controls the fluent unfolding in time of a sequence at production. An inhibition of return mechanism switches sequential elements off a fixed time after their activation. By varying the speed of movement generation through a graded "go" signal, the authors (Grossberg & Pearson, 2008) modeled variable timing of sequential arm movements including anticipatory preparation to the up-coming action. Unexpected variation of the time needed to terminate an action is not addressed. These models possibly account for a different sequential mechanism than the embodied DFT model.

2.1.5 Neural dynamic models

In this Section, I review several models that also use neural dynamics to generate sequences. These architectures are most closely related to the DFT sequence generation model presented in this thesis, therefore I describe some of them in detail. The concept of creating stable states at each stage of a sequence that lose stability in a bifurcation sets the DFT approach to sequence generation apart from other neural dynamics models (Beiser & Houk, 1998; Deco & Rolls, 2005; Rabinovich *et al.*, 2006; Selinger *et al.*, 2003). These models, described in detail below, share with my approach the commitment to principles of neural processing. They generate sequences of semi-stable states through carefully designed transient dynamics. The timing of the sequential transitions is controlled, however, by the internal dynamical properties of the neurons and thus resides at the neural time-scale. Constraints that arise from acting out the behavioral sequences by embodied agents in time-varying environments are not addressed. Although mechanisms have been proposed that aim at closing the gap between neural and behavioral time-scales (Maass *et al.*, 2002; Melamed *et al.*, 2004), these mechanisms do not address the flexibility in timing that an embodied agent faces. Several subsections that follow describe the neural dynamic approaches to sequence generation, highlighting their disembodied character.

2.1 State of the art in modeling sequence generation

Attractor networks. In a dynamical approach to motor control (Stringer *et al.*, 2004, 2007), a model has been introduced that is able to learn arbitrary motor sequences (Stringer *et al.*, 2003). This framework uses continuous attractors networks, that are related to dynamic fields. By learning associations between state and motor networks, the model can perform sequences of motor commands without feedback control, but with variable speed and force of the resulting movements. In a more cognitive setting, such a model must be embedded in a larger architecture to let the environmental input impact on an unfolding sequence.

In (Nowotny & Rabinovich, 2003), a computational mechanism is suggested, that allows to store, retrieve, and predict temporal sequences in a network of integrate-and-fire neurons connected through plastic synapses. After conditioning through repeated input of a limited number of temporal sequences, the system is able to complete the temporal sequence upon receiving the input of a fraction of them. The model is a simplified but biologically inspired architecture that allows to study the influence of several parameters, such as noise, system size, and entrainment time on the success of sequence learning.

In this neural network model, a number of input nodes provides a sequence of spikes with a fixed timing. Each input spike causes spiking of one of the integrate-and-fire neurons in a pool with all-to-all connectivity. The strength of the synapses between the neurons in the pool is changed according to an STDP rule. The activity in the neural pool is inhibited by a neuron with a slow calcium dynamics.

This work shows how synaptic plasticity enables storing time sequences of excitation of neurons as patterns of strengthened synapses. Although addressing the sequencing of bird songs and overlearned chains of motor commands, this mechanism does not have the stability property that we believe to be essential for the embodied generation of sequences. As a result, the timing of the sequential patterns is fixed by the decay constants of the dynamics (see Figure 2.1 for the time-courses of the neurons' activity during sequence retrieval). This dynamic mechanism is therefore not suited to explain sequence generation at the behavioral level, nor can it be used to organize robotic action.

Sequential decision making. A dynamic model for sequential decision making of (Rabinovich *et al.*, 2006) is formulated for variables that abstract from the activation

2. BACKGROUND AND METHODS

of single neurons. The authors discuss the stability vs. flexibility trade-off in their work. To them, stability means reproducibility of a particular sequence and resistance of the whole sequential pattern against noise. The authors design a dynamics that can produce a sequence of decision-making events. Their model includes ordinary differential equations for the dynamics of cognitive states and equations for control parameters given by decision-making rules.

The working regime of the system is a stimulus dependent competition without winner until the system reaches the "end of life" (a stable equilibrium). Before the system reaches this point the different cognitive states become "winners" just for a short time. This "winnerless competition principle" for the reproducible transient dynamics of neural systems is ensured by a non-symmetric inhibition. The parameters that control the cognitive-state dynamics are governed by a gradient descent dynamics. As the dynamics of these parameters can be made fast, the authors assume that even in

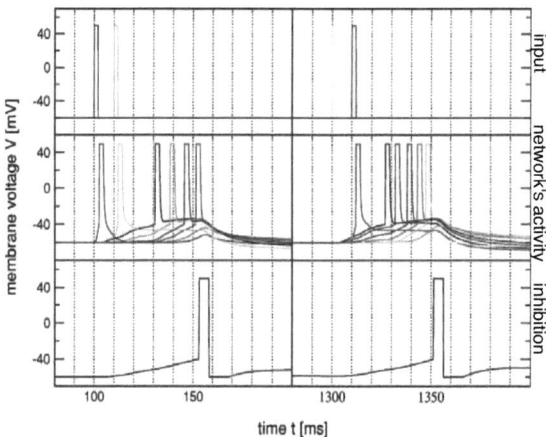

Figure 2.1: A typical recall episode for a sequential pattern in an STDP neural network. Note the fixed duration of the activity peaks. Top: input pattern. Middle: activity of the network. Bottom: a global inhibitory signal that resets the network's activity. From (Nowotny & Rabinovich, 2003).

2.1 State of the art in modeling sequence generation

a continuously changing environment, the control parameters may change in a discrete way. Thus, although the authors use continuous dynamics, they side-step the problem of linking this continuous dynamics to continuous input form the real world.

By analyzing the dynamics of their model, the authors find conditions under which the system evolves until it reaches a transient state, from which a decision emerges describing the further evolution of the dynamics. This allows for reproducibility of a sequence of transient states. However, it is far from trivial to couple this model to real sensors or guide real behavior. Moreover, the model doesn't address the problem of variable action durations in a behavioral scenario. In Figure 2.2, the time-course of the model's activity is shown. It is obvious, that the fixed timing of the trajectories depends on the parameters of the dynamics.

A similar dynamics is implemented in (Selinger *et al.*, 2003) in a two-layer architecture. Here, the neurons in a perceptual layer without lateral interactions are activated by external stimuli and, at retrieval, by so called "principle" neurons from the second layer. The "principle" neurons interact according to the dynamics of winnerless competition.

The authors recognize the problem of variable action durations and arrive at a solution that contradicts their initial modeling hypothesis that a sequencing architecture must be a transient dynamics. The authors discard attractor dynamics as unsuitable for the description of reproducible sequential neural dynamics (Afraimovich *et al.*, 2004; Rabinovich *et al.*, 2008). Although they suggested a solution to this problem, they do not persue it and do not try to connect their model to real actuators and sensors to see whether they can explicitly specify all the components and control signals for their dynamics in a real-world implementation.

Detailed neural dynamics. Deco and Rolls (Deco & Rolls, 2005) present another neural dynamic architecture capable of generating sequences. In this dynamics, each item in a sequence corresponds to an active pool of neurons. This state of the network is an attractor state of the dynamics and the neurons fire until an external inhibitory signal quiesces the network. Partial adaptation happens in a neuronal pool for the active neurons. Thus the item that was presented, and thus activated, earlier will be less inhibited due to the adaptation and will be activated when the inhibition is released. In this manner, the network reproduces a sequence of states presented to

2. BACKGROUND AND METHODS

it during learning. The authors provide three different molecular accounts for the postulated adaptation mechanism at the level of neuro-transmitters.

The architecture consists of two excitatory and one inhibitory pools of neurons. One of the excitatory pools is specific and encodes, for example, the identity of a visual object, or a phoneme, or a simple motor action to be remembered as a part of a temporal sequence. The remaining excitatory neurons are in a nonselective pool, which introduces some noise to the simulations. The inhibitory pool serves for the global competition in the network.

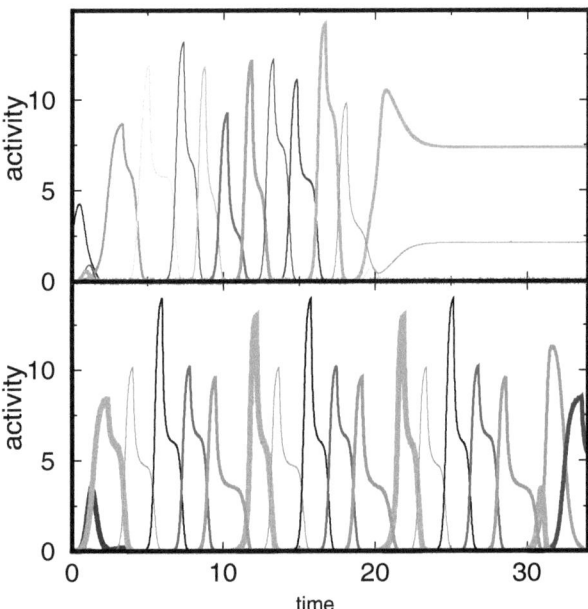

Figure 2.2: A decision-making dynamics (Rabinovich *et al.*, 2006). Top: a typical output of the model. Bottom: An example of the repetitive decisions that may be observed in a repetitive environment with a small percentage of uncertainty. Note the fixed timing of the sequential decisions: the peaks have similar durations.

2.1 State of the art in modeling sequence generation

The authors use leaky integrate-and-fire neurons for modeling the excitatory neurons and the inhibitory interneurons. The synaptic inputs to an integrate-and-fire neuron are basically described by a capacitor connected in parallel with a resistor through which currents are injected into the neuron. These current injections produce excitatory or inhibitory postsynaptic potentials. These potentials are integrated by the cell, and if a threshold is reached, a pulse (spike) is fired and transmitted to other neurons, and the potential of the neuron is reset. The incoming presynaptic pulse current from another neuron is low-pass filtered by the synaptic and membrane time constants.

This provides a mechanism that implements a temporal memory of the previously activated pool. When the attractor state of the network is shut down by the inhibitory input, the attractor state that subsequently emerges when firing starts again will be different from the state that has just been present because of adaptation of the synaptic or neuronal processes that supported the previous attractor state. To assure that one of the specific pools is active, and to promote high competition between the possible specific pools (the items in the sequence), all specific pools are continuously excited externally with the same nonspecific input by increasing the external input's firing rate impinging on the excitatory pools of the network. The function of this external input is thus to maintain as selected the set of items in the sequence being remembered.

To show how the model can implement the learning of temporal sequences of items, the authors perform simulations of how temporal order of a two- or three-element sequence is maintained. Figure 2.3 shows the longer of the produced sequences. Note, that as the amount of adaptation is critical here to produce the correct sequence of items, the variability of action durations is limited, although the items in the sequence are modeled as stable attractor states and the transitions are triggered by an external signal. In fact, in the sequences presented by the authors, the action durations are identical for all sequences and the possibility of variable action durations is not discussed. Moreover, the complexity of the underlying dynamics seems to keep the authors from implementing longer sequences. Thus, this neural dynamic mechanism might account for the read-out of short sequences of neural activity, which might be relevant for neural processing, but cannot account for overt sequential behavior.

In summary, the biologically detailed dynamical models of sequence generation reviewed do provide insights into neural mechanisms of sequential action but do not address the question of how these neural mechanisms drive overt sequential behavior of

2. BACKGROUND AND METHODS

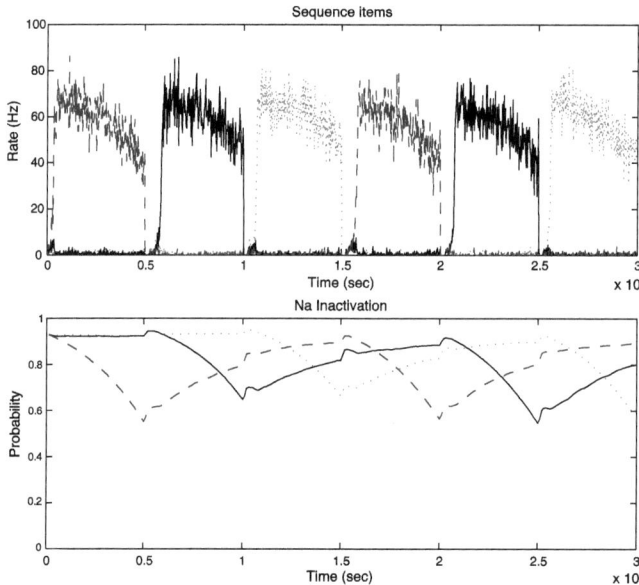

Figure 2.3: Output of the sequence generating architecture of (Deco & Rolls, 2005). Top: rate of average activity in three pools of simulated neurons that correspond to three items in a sequence. Bottom: The probability of spiking for these neural pools.

an agent acting in a physical environment, which the agent perceives through it's own sensors and in which the agent acts autonomously.

Hebbian cell ensembles. A conceptual approach to neural dynamics that is closely related to DFT is based on the idea of Hebbian cell assemblies (Wennekers, 2006; Wennekers & Palm, 2007). In this framework, ensembles of dynamical neurons are effectively bound by strong excitatory interaction in the presence of global inhibitory coupling. Like DFT, these approaches account for stable patterns of population activity. The neural field concept has been shown to emerge as the limit case of homogeneous coupling structure (Deco et al., 2008; Potthast & beim Graben, 2009). Within the framework of Hebbian cell assemblies, a model of sequence generation has been pro-

2.1 State of the art in modeling sequence generation

posed that shares key ideas with the DFT architecture. In that model, sequences may be generated transiently from semi-stable states of neuronal assemblies ("autonomous mode"), but may also emerge from actual attractor states. The attractors can be switched by a global signal, similar to the condition of satisfaction signal of the DFT sequence generation model (Wennekers & Palm, 2009). This global signal is excitatory, however, whereas the condition of satisfaction signal is inhibitory, which guarantees robust switching even in the face of sensory noise, fluctuating input, and symmetry in the system. The Hebbian cell ensembles sequence generation model is not designed to be coupled to real-world sensors or to control a behavior.

2.1.6 Sequence generation in robotics

Naturally, the problem of embodiment is addressed within the domain of autonomous robotics and autonomous agents research. Here, sequence generation is more typically looked at as a problem in behavioral organization, in which rules determine the sequential ordering of actions (Steinhage & Schöner, 1998). From that perspective, serial order is the simplest case, in which any order is possible (although serial order includes the problem of learning arbitrary sequential arrangements). A typical approach is to represent actions as discrete nodes at one or a number of hierarchical levels. Directed links between these nodes define the logical structure of behavioral organization, whereas inputs represent goals and environmental conditions. Such architectures have demonstrated behavioral sequencing in simulated environments and simple robotic demonstrations (Maes, 1989; Payton *et al.*, 1990; Tyrrell, 1993). In order to apply these models in the robotic domain, the temporal continuity of the actions and their finite duration must be taken into account. One way to do that is to use event-driven processing in finite state machines which are in stationary states between transitions (Arkin & MacKenzie, 1994; Kosecka & Bajcsy, 1993). In this view, cognitive architectures are effectively decoupled from the sensory systems. The downside is that the stability problem is shifted to the level of sensory preprocessing, at which the relevant problems of sensor fusion, salient events detection, and segmentation of relevant objects are still broadly unsolved. Because many of these processes are actually cognitive in nature, their relegation to a system outside behavioral organization may be problematic.

2. BACKGROUND AND METHODS

We thus argue for the necessity, also for robotic applications, of a coherent theoretical framework, that links perceptual processes, motor behavior and cognitive control, which includes process models of embodied sequence generation.

2.1.7 Open issues in modeling sequential action and the problem statement for the thesis

Because current neuronal models of sequence generation (Botvinick & Plaut, 2006; Brown *et al.*, 2000; Burgess & Hitch, 1999; Deco & Rolls, 2005; Elman, 1990; Houghton, 1990; Page & Norris, 1998) address the constraints of embodied sequence generation only partially, there is, to my knowledge, no comprehensive theoretical account for embodied and situated sequence generation that explains how behavioral sequences are acquired and produced in the real world.

The goal of this thesis is to establish such an account. By employing the framework of Dynamic Field Theory (DFT) we ensure that this approach is neuronally based and consistent with the two constraints of *time-continuous* processing on *graded* representations (Schöner, 2008). The units of representation in dynamic fields are peaks of activation defined over continuous behavioral dimensions. Such peaks emerge *autonomously* from the interplay of neuronal interaction and sensory input and may be localized as dictated by current input or also be determined by memory traces acquired during learning. As a result, dynamic fields are capable of both representing categorical as well as *continuous metric* information. Because the peaks are attractor states of the underlying neuronal dynamics, they may be sustained over unpredictable delays enabling *flexibly timed* actions.

2.2 Dynamic Field Theory to cognition

2.2.1 Dynamic equation of a neural field

Dynamic Field Theory (DFT) is a phenomenological description of the dynamics of neuronal activation that abstracts from the discreteness of neurons as units of neuronal networks and from biophysical mechanism of spiking (Amari, 1977; Ermentrout, 1998; Wilson & Cowan, 1973). As a theoretical language, DFT has been widely used to model motor, perceptual, and cognitive functions (Schöner, 2008). The central idea of DFT

2.2 Dynamic Field Theory to cognition

is that neural representations are defined over continuous dimensions such as the direction and amplitude of a reaching movement or the retinal location or color of a visual stimulus. These dimensions span a conceptual motor or feature space, over which an activation function is defined. For each location, x, along these dimensions, the activation function, $u(x)$, indicates the presence of information by high levels of activation and the absence of information by low levels of activation. High levels of activation imply that neuronal representations or effector systems onto which the activation field projects are impacted by the activation variable. Low levels imply that no such impact takes place. The activation function, $u(x,t)$, evolves in time, t, as described by the integro-differential equation (Amari, 1977):

$$\tau \dot{u}(x,t) = -u(x,t) + h + \int f(u(x',t))\omega(x-x')dx' + S(x,t). \quad (2.1)$$

Here, τ is a time constant of the dynamics that determines how quickly the activation function, $u(x,t)$, relaxes to an attractor state that emerges from the stabilization factor $-u(x,t)$ and the additive contributions: the negative resting level, $h < 0$, the lateral interactions shaped by the kernel, $\omega(x-x')$, and the external input, $S(x,t)$. The kernel is a bell-shaped function containing both excitatory connectivity of strength c_{exc} over the range, σ, and inhibitory connectivity of strength c_{inh} over all ranges (2.2) (in implementation, periodic boundary conditions are used to perform convolution).

$$\omega(x-x') = c_{exc} \exp\left[-\frac{(x-x')^2}{2\sigma^2}\right] - c_{inh} \quad (2.2)$$

The homogeneity of the connectivity pattern is characteristic for dynamical neural fields and, together with the bell-shaped kernel stabilizes a localized peak of activation as attractor solution. This requires a non-linearity, $f(u(x,t))$, that thresholds the output of the field's dynamics as a soft sigmoid:

$$f(u(x,t)) = \frac{1}{1+\exp[-\beta x]}. \quad (2.3)$$

2.2.2 Dynamic regimes and instabilities of a dynamic neural field

The field dynamics has different dynamic regimes. These can be analyzed in some instances using analytical techniques (Amari, 1977; Ermentrout, 1998; Wennekers, 2002), which is one of my motivations for using this particular formulation.

2. BACKGROUND AND METHODS

In the absence of external input, one attractor state has activity in the field close to the negative resting level, h. This sub-threshold solution remains stable when weak localized input, $S(x,t)$, is introduced as long as the summed activation level, $h+S(x,t)$, does not surpass levels at which the lateral interaction becomes engaged. When that threshold is passed, the output, $f(u(x,t))$, of the interaction function drives the system through the *detection* instability, in which the sub-threshold solution disappears. Activation grows near the field sites at which localized external input was largest, developing into a localized peak of activation that inhibits the activation field elsewhere. The peak solution is stabilized against decay by the local excitatory interaction and against lateral diffusion by global inhibitory interaction (Figure 2.4). Such peaks thus form the second category of attractor states, which are self-stabilized by intra-field interactions, but also track changing localized input. Self-stabilized, localized activation peaks are the units of representation in DFT. For instance, objects in the visual field may be represented as peaks in a field spanned by visual features and visual space. Action plans or motor primitives may be represented by peaks in fields spanned by movement parameters. In DFT, all motor, perceptual, or cognitive states are represented by such attractor states, so that they are resistant to variations in input or coupling across different fields.

The sub-threshold solution may be inhomogeneous with various localized small

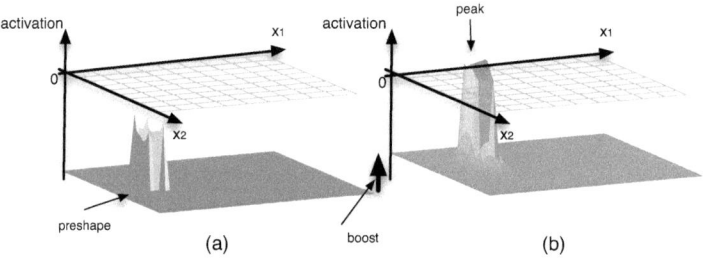

Figure 2.4: Detection instability in a two-dimensional dynamic neural field.

2.2 Dynamic Field Theory to cognition

bumps of activation due to weak input that represents learned categories within a feature dimension. Such inhomogeneities may be amplified into a self-stabilized peak by pushing the field through the detection instability by a rise of the resting level that provides a "boost" to the overall level of activation.

When localized input is removed, the detection instability may be experienced in reverse. In that *forgetting* instability, the self-stabilized peak becomes unstable and the system relaxes to the sub-threshold attractor. The forgetting instability happens at lower levels of localized input than the detection instability, so that there is a bistable regime in which detection decisions are stabilized. For sufficiently large resting levels, h, or strong lateral excitation in the neural field, the forgetting instability may not occur even when localized input is removed entirely. In this case, the self-stabilized peak is sustained in the absence of the localized input that first induced it. The forgetting instability may be induced, however, by lowering the resting level.

The two instabilities translate graded changes in a neural field's input into macroscopic decisions represented by bifurcations in the field dynamics. These decisions are elementary forms of embodied cognitive function. Through these instabilities, dynamic fields provide an interface between the sensory and motor surfaces and the associated sensory-motor processes on the one hand, and a cognitive dynamics on the other hand, that enables agents to pursue goals that are not at all times dependent on sensory input.

2.2.3 Dynamics of discrete neural nodes.

Localized activity peaks are units of representation in the DFT. In a given environment, an activity peak may emerge at the same location, or a discrete number of locations in a dynamic neural field. If a memory trace that builds-up at those locations is strong enough, the field reacts categorically to fluctuating perceptual inputs, i.e. peaks emerge at locations of the long-term memory traces, or preshape, even if the location within the field of the perceptual input that brings the field above threshold does not match the location of preshape precisely. In this case, one might use several discrete nodes, or even a singe node in the limit case of a single possible location, to represent the activity in the field. Activity of these nodes follows a dynamics that is similar to the dynamics

2. BACKGROUND AND METHODS

of a neural field, Eq. 2.4.

$$\tau \dot{u}(t) = -u(t) + h + c_{exc} f(u(t)) + S(t). \tag{2.4}$$

Notation here is analogous to that of Eq. 2.1. The self-excitatory term of strength, c_{exc}, provides for a bi-stable dynamics: when the node is pushed above the activation threshold of the output function, $f(\cdot)$, in a detection instability, the positive activity of the node is stabilized and can sustain fluctuations of the inputs signal, $S(t)$. At sufficiently strong self-excitation, the node will stay active even if the input that has pushed it through the threshold ceases (memory instability). An inhibitory input is required then to bring the node into an inactive state (forgetting instability).

In the architectures presented in this thesis, I will use several sets of discrete dynamical nodes to represent units that follow the Amari dynamics but are not necessarily embedded in a metrical space.

2.2.4 Coupling between neural fields and nodes

In DFT, the dynamic neural field spends most of the time in an attractor, either a sub-threshold, inactivated one, or an activated one with one or several activation peaks. Transitions between these states are fast and infrequent. Stability of the attractor states allows not only to filter out noise and sustain a certain state in the face of external perturbations, it also enables coupling between neural fields that keeps the qualitative dynamics invariant. Coupling between neural fields can be parametrized to be weak and then it does not remove or qualitatively alter the solution and does not lead to an unpredictable behavior. The states of neural fields in the architecture remain stable. This invariance property enables building up architectures of neural fields from individual field components, each designed to be in a chosen dynamic regime.

Couplings between neural fields is modeled as synaptic connections, or "maps". Such maps may project activation of discrete nodes onto a dynamic field or transfer activation between fields of several dimensions along a shared dimension. Couplings between fields of different dimensionality are discussed in the context of a scene representation architecture (Zibner et al., accepted). The connection weights that couple different dynamical neural fields and nodes may be learned in a Hebbian learning process: the weights are strengthened that connect two simultaneously active regions. I use this learning mechanism in the DFT sequence generation model presented in Chapter 5.

2.2.5 Problem of sequentiality in DFT to cognition

The properties of dynamic neural fields enable building up architectures that model cognitive behavior. The behavior of the architectures can then be compared to results of empirical studies of cognitive behavior of adults, children, and infants. Because of the graded metrical representations of DFT and the temporally continuous dynamics, a large number of effects can be understood in this frameworks. Models of visual working memory (Johnson *et al.*, 2009) and visual cognition (Johnson *et al.*, 2008), spatial working memory (Schutte *et al.*, 2003; Spencer & Schöner, 2006), spatial language and cognition (Lipinski *et al.*, 2006, 2009e), perseveration (Spencer *et al.*, 2001), and infant looking (Perone *et al.*, 2007) are examples of this line of research. On the other hand, robotic architectures can also be developed within DFT: an object recognition system (Faubel & Schöner, 2008), a scene representation (Zibner *et al.*, 2010), and a spatial language architecture (Lipinski *et al.*, 2009d) address different aspects of robotic action and were tested in different robotic scenarios. In both cognitive science and robotic implementations, the property of stability of neural fields enables creating of the coupled dynamic fields architectures as well as coupling to sensory input. The property of stability, however, stays in conflict with sequentiality: a mechanism to turn-off an attractor state autonomously or through a signal that is autonomously extracted from the sensory stream has not been introduced so far. This has limited the autonomy of the created architectures. The sequentiality has also to be introduced to arrive at a solution to the problem of behavior organization within robotics research and the problem of executive control in the cognitive science research. The models presented in this thesis introduce a mechanism that enables sequence generation within DFT and is, therefore, the first step towards such autonomous architectures that enable complex, flexible and robust behavior.

2. BACKGROUND AND METHODS

3

The DFT sequence generation architecture. A one-dimensional model.

3.1 The task setting

Generating ordered sequences of actions has different facets. In this thesis I focus on the ordinal component of it, thus setting the goal to develop a framework in which sequences of stable states can be generated, with transitions between them controlled by an external signal. In this chapter I present a DFT model for generating homogeneous sequences of action in which each action can be described by the same characteristic dimension. Over that dimension, a dynamic neural field is defined. This simplification makes it possible to separate the problem of sequentiality from the problem of coordinating actions across different modalities, which might compete for resources, exclude or enforce simultaneous execution, and facilitate or prohibit certain orders (i.e. behavior organization). Extending the model to multiple modalities is considered in Chapter 6.

In a robotic implementation of the model, a color-search task serves to demonstrate the properties of the neural dynamics. For this task, a single dimension – the color of an object that the robot must search – is sufficient to describe the robotic actions. The color-search action at each stage of a sequence takes various amounts of time, so this simplest setting probes the capacity to reconcile stability (stick to the task until

3. THE DFT SEQUENCE GENERATION ARCHITECTURE. A ONE-DIMENSIONAL MODEL.

solved) and switching (search for the next color once done).

This chapter is structured as follows. An intuitive description of how this DFT architecture works is provided in Subsections 3.2.1, 3.2.3, 3.2.4. The mathematical description is presented in Section 3.3, followed by the particular robotic implementation in Section 3.4. Results of robotic demonstrations are presented in Section 3.5 and are discussed in Section 3.6. Work presented in this chapter was published in (Sandamirskaya & Schöner, 2008).

3.2 The DFT architecture for sequence generation

At the core of the dynamic field model of sequence generation lies a stack of neuronal activation fields (Figure 3.1). These fields are spanned over the same neural dimension that encodes a parameter relevant for the particular behavior. Each layer in the stack represents a particular ordinal position in the sequence. The directed links between these *ordinal fields* control the sequential activation of localized peaks in the ordinal stack. The location of each peak in the ordinal stack is coded by preshape – a memory trace laid down during sequence learning. Activity in the ordinal stack is integrated in an *output field* defined over the same dimension. Positive activity in the output field controls the action system, ultimately guiding the behavior of an agent. The output field also determines the pattern of sensory information that signals the completion of the current action providing input to a neuronal representation of a *condition of satisfaction* (Searle, 1983). The condition of satisfaction signal triggers a transition to the next ordinal position and is deactivated during the transition.

3.2.1 Ordinal fields

The stack of neural fields illustrated in Figure 3.1 encodes sequential order in that, aside from brief transitions, only one ordinal layer at a time can have a self-stabilized activity peak. The location of the active field in the stack encodes the ordinal position of the corresponding action in the sequence. The location of the peak within the active field encodes the content of that action.

Activity in an ordinal layer homogeneously inhibits the predecessor layer and provides excitatory input to the successor layer. The excitatory coupling is shunted, however, by the condition of satisfaction system. When that system is active, the excitatory

3.2 The DFT architecture for sequence generation

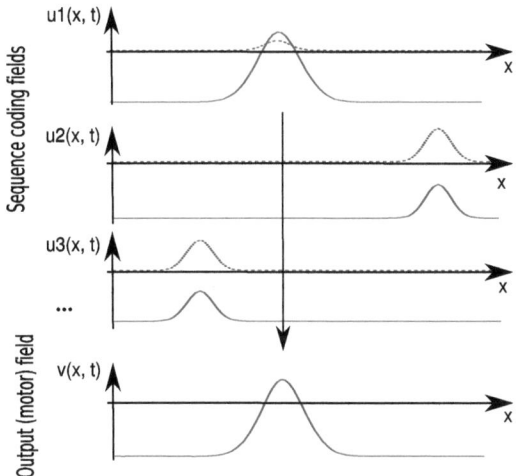

Figure 3.1: Overview of the DFT architecture for generating action sequences. Red lines show activation of neural fields, blue lines show the memory traces from which activity peaks may be induced.

3. THE DFT SEQUENCE GENERATION ARCHITECTURE. A ONE-DIMENSIONAL MODEL.

coupling is effective for the ordinal layer, in which the activity peak overlaps with the peak in the output layer. The activity is then propagated in the ordinal stack. The peak induced in the successor layer suppresses activity in the predecessor layer. In the transition phase, when two ordinal fields are active, the condition of satisfaction (CoS) system, introduced further, is inhibited.

Where in an ordinal field the activity peak is located is defined by a memory trace, layed down during sequence learning.

3.2.2 Sequence memory: preshape of the ordinal fields

Laying down a memory trace of self-stabilized peaks of activation is a simple form of learning in DFT (Erlhagen & Schöner, 2002). The memory trace preshapes the activation field. It means that a localized peak can be induced from the memory trace, or preshape, by a homogeneous boost.

Based on this mechanism, sequence learning is realized by accumulating memory traces during a learning session, when a sequence of actions is demonstrated to the agent by inducing a sequence of the corresponding activity peaks through its sensory system. The memory trace can also be understood as a form of Hebbian learning that establishes connection weights between the fields in the ordinal stack and higher-level nodes that encode sequences in their entirety (top left of Figure 3.3). To enact a particular sequence, the corresponding higher-level node is activated and the fields in the ordinal stack are preshaped at the locations determined by the memory traces.

3.2.3 Output field

The activity in the stack of ordinal fields is integrated in the output field. The resulting stable activity peak controls action of the agent, until the switch to the next steps has been achieved. The output layer thus stably represents the current action throughout the variable time interval that the physical realization of this action takes.

Figure 3.2 illustrates four stages in the dynamics of the ordinal stack and output field: (a) when a sequence is initiated, (b) the first action is acted out, (c) the transition to the second action happens and (d) the second action is acted out.

3.2 The DFT architecture for sequence generation

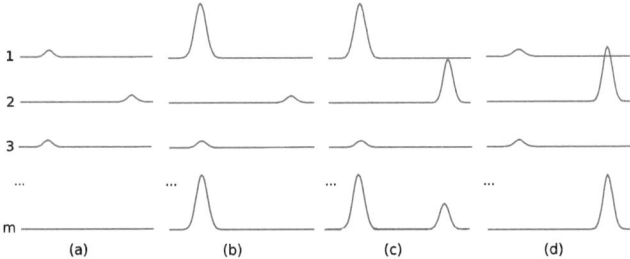

Figure 3.2: Four stages of the sequence generation dynamics. (a) The ordinal fields (1, 2, 3, ...) are preshaped at locations that correspond to actions in the memorized sequence. (b) A homogeneous boost to the first ordinal field starts sequence generation. A localized peak of activation is induced at the location specified by the memory trace in the first ordinal field. A matching peak is induced in the output field. (c) The condition of satisfaction signal releases shunting of the excitatory connection between the ordinal fields. Activation in the first ordinal layer induces a peak in the second ordinal layer at the preshaped location. (d) The newly established peak in the second ordinal layer inhibits the original peak in the first ordinal layer. The output field carries a peak at a matching location.

3.2.4 Condition of satisfaction system

A fundamental conflict of sequence generation is between the need to stabilize the behavioral state at a given step in the sequence and the need to destabilize that state in order to switch to the next step of the sequence. In the DFT architecture, this transition is mediated by a condition of satisfaction system. This is a neuronal dynamics with the same bi-stability between "on" and "off" states as the neuronal field. In fact, the single neuron we are using could be viewed as the activation in a peak of a neuronal field, which could accommodate a range of conditions of satisfaction nodes needed in more complex scenarios.

The sensory signal about accomplishment of an action is generated from sensory input, the strength and duration of which fluctuate (bottom right of Figure 3.9). To stabilize the condition of satisfaction signal, the neuronal dynamics is bistable, so that the neuron remains "on" even if the sensory signal drops below the initial threshold. Only when the neuron is actively inhibited by negative input does it return to the "off" state. Such negative input comes from the stack of ordinal fields when it is in

3. THE DFT SEQUENCE GENERATION ARCHITECTURE. A ONE-DIMENSIONAL MODEL.

transition between two states: the input function detects correlation between suprathreshold activation in an ordinal field and in the output field, observed simultaneously for two consecutive ordinal layers.

The condition of satisfaction neuron shunts excitatory coupling from any ordinal field to its successor. Thus, the condition of satisfaction system does not need to "know" at which ordinal position the sequence currently is.

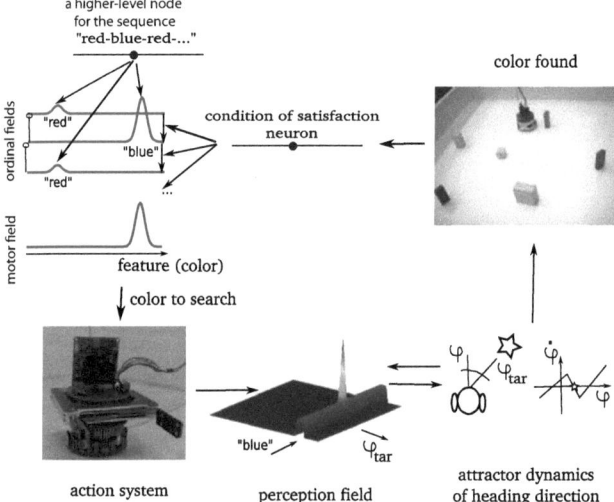

Figure 3.3: The robotic implementation of the DFT sequence generation architecture for the color-search task.

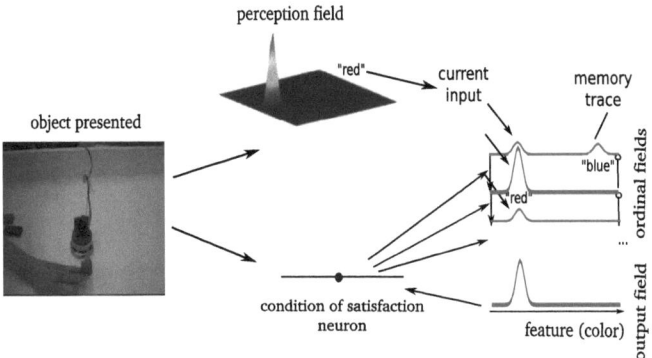

Figure 3.4: The DFT sequence generation architecture in a learning session of the color-search task.

3.3 Mathematical description of the model

The dynamics of each field, $u_i(x,t)$, in the ordinal stack (ordinal index, $i = 1, \ldots, N_{Sc}$) is

$$\tau_{Sc}\dot{u}_i(t,x) = -u_i(t,x) + h + \int f(u_i(t,x'))w_{Sc,Sc}(x,x')dx'$$
$$+ C_+ f(\xi_{cs}) \int f(v(x',t))f(u_{i-1}(x',t))dx' \qquad (3.1)$$
$$- C_- \int f(u_{i+1}(x',t))dx' + p_i(x,y,t)$$

with similar parameters as the generic field equation Eq. 2.1. The constants, C_+ and C_- control the boosting and deboosting coupling, the preshaping input, $P_{iY}(x,t)$ comes from the higher level neuron Y. Sensory feedback about action completion comes through the condition of satisfaction node, ξ_{cs}.

The dynamics of the motor field, $v(x,t)$, is:

$$\tau_M \dot{v}(x,t) = -v(x,t) + h + \int f(v(x',t))w_{MM}(x,x')dx'$$
$$+ \Sigma_{i=0}^{N_{Sc}} \left[\int f(u_i(x',t))w_{MSc}(x,x')dx' + C_+ \right] \qquad (3.2)$$

with analogous notation.

3. THE DFT SEQUENCE GENERATION ARCHITECTURE. A ONE-DIMENSIONAL MODEL.

The condition of satisfaction neural dynamics is

$$\dot{\xi}_{cs}(t) = -\lambda \xi_{cs}(t) + h_{cs} + \mu \; f(\xi_{cs}(t)) + I(t) + F(t) \tag{3.3}$$

where λ and μ are constants, h_{cs} is the resting level, and $f(\cdot)$ is the sigmoidal non-linearity, which provides self-excitation to the neuron, ξ_{cs}. The sensory signal, $I(t)$, is obtained from the vision system when the condition of satisfaction for the particular action is detected. The negative reset signal, $F(t)$, is obtained from the ordinal stack when two subsequent ordinal fields are active in the transition phase:

$$F(t) = c \sum_i \int f(u_i(x,t)) dx \int f(u_{i-1}(x,t)) dx \tag{3.4}$$

The memory trace mechanism for learning a sequence is implemented as the following dynamics:

$$\begin{aligned}\tau_p \dot{p}_i(x,y,t) = \lambda_{build} f\big(u_i(x,t)\big) \cdot \Big(&- p_i(x,y,t) \\ + f\big(u_i(x,t)\big) \cdot f\big(U(y,t)\big)\Big),\end{aligned} \tag{3.5}$$

where $p_i(x,y,t)$ is a synaptic connection from the site y in a higher level pool of neurons $U(y,t)$, to the site x of a sequence coding field $u_i(x,t)$. τ_p is the time constant, λ_{build} is the rate of strengthening of connections driven by simultaneous activation of $u_i(x,t)$ and $U(y,t)$.

3.4 Robotic implementation of the DFT sequence generation architecture

To demonstrate the core properties of the DFT model for sequence generation (autonomous sequence learning and production, flexible timing and guiding a real-world action) we implement the model on a mobile robot. The simple robotic scenario was a color-search task, in which a sequence of colors is presented to the robot. That sequence is autonomously extracted from the visual stream (Figure 3.5). The robot is then put in an arena in which a set of colored blocks is distributed (Figure 3.6). The robot drives around the arena and avoids obstacles. Its task is to visit the colored blocks in the memorized serial order.

For this robotic demonstration the action can be characterized by the color that the robot must search for at each stage in the sequence. Thus, the dimension over

3.4 Robotic implementation of the DFT sequence generation architecture

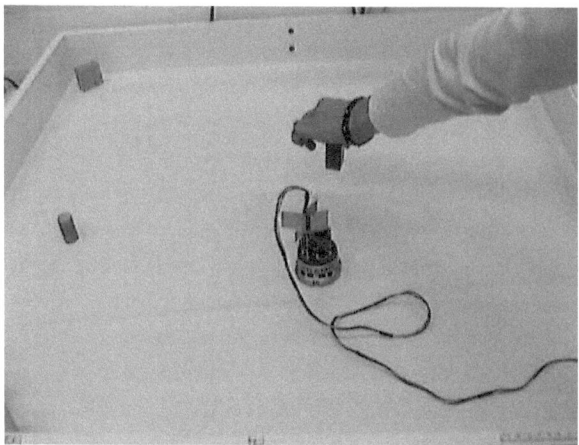

Figure 3.5: Learning of a sequence of colors. The user shows color blocks to the robot in a given serial order.

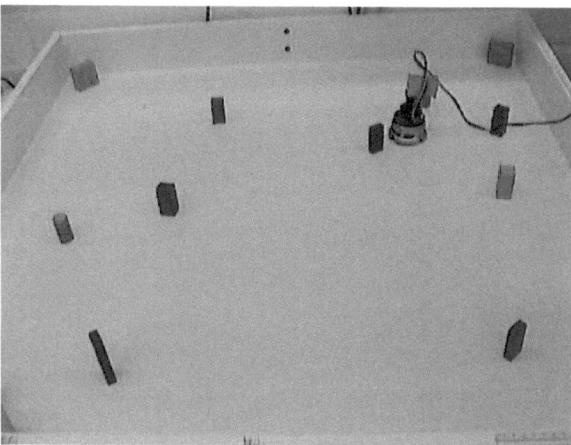

Figure 3.6: Generation of a color-search action sequence in the robotic arena.

which the ordinal and output (motor) field are defined is the hue value of the HSV (hue-saturation-value) color scheme.

3. THE DFT SEQUENCE GENERATION ARCHITECTURE. A ONE-DIMENSIONAL MODEL.

An activity peak in the output field thus specifies the range of hue values relevant at the particular stage in the sequence. The output field provides input to a two-dimensional perceptual color-space field, defined over the dimension color and the dimension of the horizontal axis of the image plane (Figure 3.7). The perception field is a two-dimensional dynamic neural field that associates the color of visual targets with the heading direction in which the targets are seen. It receives two-dimensional input extracted from the camera image: A histogram of hue values obtained within each column of the camera image defines the input function along the color dimension at the heading direction, into which this image column is pointing. A peak of activation in the output layer of the sequence generating system provides a ridge of input across all heading directions, which effectively boosts all those parts of the visual array, at which objects are seen with a color that matches the current state of the output field. Such matching input leads to a self-stabilized peak in the two-dimensional field. If there are several candidate objects in the visual array, the two-dimensional perception field selects one (typically the largest object) and then stabilizes that selection decision through the neural field dynamics

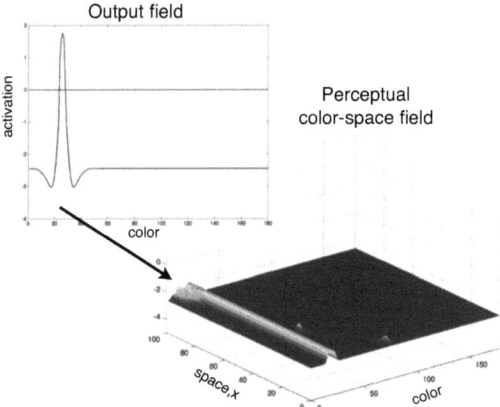

Figure 3.7: Ridge-shape input from the output field of the DFT sequence generation architecture to a two-dimensional perceptual color-space field.

3.4 Robotic implementation of the DFT sequence generation architecture

During learning, the teacher shows colored blocks to the robotic camera.

The condition of satisfaction node receives sensory input from the robot's vision system. The count of pixels whose current color matches the color specified by the output field cues when the planned action has been achieved. When this input reaches a critical level, the "off" state of the condition of satisfaction node becomes unstable and the condition of satisfaction node switches to "on".

The robot is controlled by a dynamics of heading direction which integrates two kinds of contributions (Schöner et al., 1995). The position of an activation peak in the perceptual field along the axis of heading direction controls an attractive force-let. Distance signals obtained from on-board active infra-red sensors modulate the strength of repellors that implement obstacle avoidance. The rate of change of heading direction is input to servo-controllers on the robot vehicle, specifying the difference in velocity of the left and right active wheel so that the desired turning rate is generated. The forward velocity of the vehicle is also controlled so that the robot moves slowly when searching, when near an obstacle, or when close to a target. The velocity is faster when a target has been detected toward which the robot is moving. The forward velocity sets the mean of the signals sent to left and right wheel servo.

The combined effect of these dynamics is that the vehicle moves forward while avoiding obstacles, so that the system effectively searches its environment. As soon as the robot has detected an object with the currently requested color, it moves toward that object. The visual image of the object ultimately becomes sufficiently large within the camera plane to trigger a condition of satisfaction signal.

The perceptual field, $u_p(x, \phi)$, is a 2D variant of the Amari system, Eq. 2.1. The dynamics of heading direction, ϕ is

$$\tau_\phi \dot{\phi} = \lambda_{\text{obs}} F_{\text{obs}}(\phi) - \lambda_{\text{tar}} \int f(u_p(x, \psi))(\psi - \phi) dx d\psi \qquad (3.6)$$

The obstacle forcelet $F_{\text{obs}}(\phi)$ is a sum over the contributions of the 6 infrared sensors of the Khepera robot. Each sensor contributes with a strength that depends on the sensed distance to an obstacle, and an angular range that depends on the opening angle of the sensors, on the robot size, and on the sensed distance (see (Bicho et al., 2000) for details).

3. THE DFT SEQUENCE GENERATION ARCHITECTURE. A ONE-DIMENSIONAL MODEL.

3.5 Results of robotic demonstrations

3.5.1 Learning

First, we demonstrate one-shot learning of a sequence (Figure 3.8). Five colored bricks are presented to the robot in succession. In each case, input from the camera generates a peak in the perceptual field at the corresponding hue value and heading direction. The supra-threshold component of the peak summed along heading directions is a distribution of activation over hue space and provides input to each field in the ordinal stack. In the beginning of the learning session, the first ordinal field receives a homogenous boost that corresponds to a "go" signal. When the first color block is presented by the user, the first layer develops a peak at the location on the color dimension that is specified by the input from the perceptual field. A memory trace associated with the ordinal layer accumulates activation at the peak location. This memory trace serves as the memory for the particular item in the sequence. When the memory trace reaches a critical mass and the perceptual field registers a removal of input (because the brick has been removed from the visual array of the robot), a signal is generated by the condition of satisfaction node which now autonomously organizes the switch to the second ordinal layer. When the second brick is presented, this layer develops a peak at the associated hue value, suppressing the peak in the first layer and laying down a memory trace. This continues until all colored objects have been shown to the agent.you

3.5.2 Sequence generation

To execute the sequential search task, the robot is set into an arena in which colored bricks have been distributed (top right of Figure 3.3). The "go" signal is given, leading to activation of a peak in the first layer of the ordinal stack at the location representing the first color in the sequence and to activation of an associated peak in the output field (Figure 3.9). This peak provides a ridge input at the corresponding hue value to the two-dimensional color-heading direction perception field (bottom of Figure 3.3). The vehicle moves around the arena, avoiding obstacles. The vehicle keeps track of the heading direction by integrating its rate of change. The camera image is continuously transformed into the coordinate frame of the heading direction and input into the perceptual field. When visual input is generated that overlaps sufficiently with the preshaped ridge at the requested hue value, a peak forms in the perceptual field. This

3.5 Results of robotic demonstrations

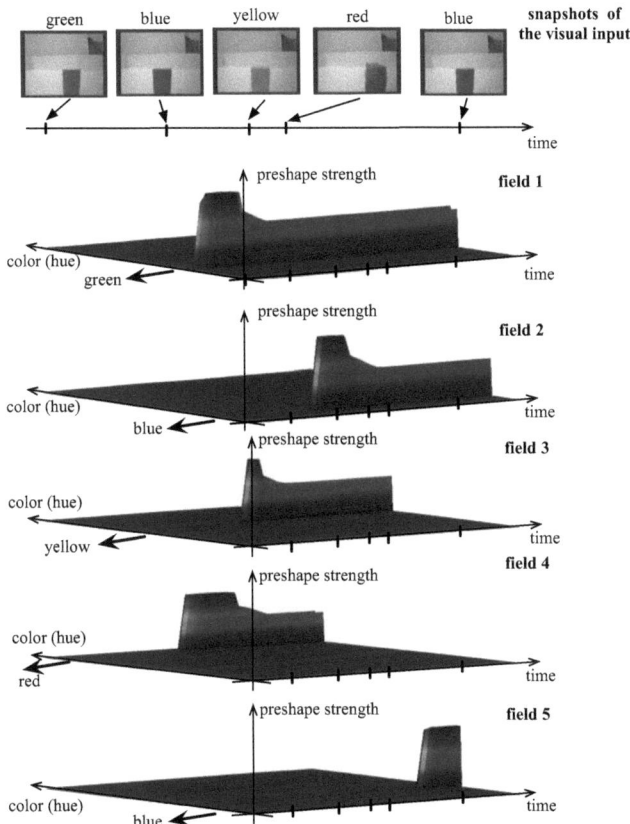

Figure 3.8: A learning session. **Top**: snapshots of the visual input: five blocks of different colors are put in front of the robot in a succession. **Bottom**: time-courses of formation of the memory traces (preshapes) in five dynamic neural fields of the ordinal stack. At the end of the learning process, the preshapes of the ordinal fields hold the memory of five colors in the order in which they were presented.

3. THE DFT SEQUENCE GENERATION ARCHITECTURE. A ONE-DIMENSIONAL MODEL.

peak defines a movement target and the vehicle moves in the required direction, while continuing to avoid obstacles. Typically, the vehicle is able to approach the object, so that the object's projection onto the visual array grows in size, until it looms sufficiently large in the camera image to send a sensory signal to the condition of satisfaction system (bottom right of Figure 3.9).

At this point, a transition in the ordinal stack is triggered. The second ordinal layer builds a peak at the second learned color value, which is replicated in the output layer. As this happens, the condition of satisfaction node switches back to the "off" state and the peak in the first ordinal layer is suppressed by inhibition from the second ordinal layer. The change of location of the peak in the output layer shifts the hue value at which a ridge is input into the perceptual field. The current peak in this field therefore decays. This removes the attractor of heading direction in the direction of the previous target and effectively puts the system back into search mode. The robot again moves around, avoiding obstacles, until it encounters in its visual array enough color information that overlaps with the new preshaped ridge, leading to the generation of a new peak in the perceptual system. This reinstates a movement target, toward which the robot navigates. As that target is approached, another sensory signal is sent to the condition of satisfaction system, leading to the second transition.

Figure 3.9 shows four such transitions as recorded from the life robot. In each case, the peak in the output layer lasts as long as it takes to reach a target of the specified color. Naturally, the time needed to find a target varies depending on the configuration of the robot and the environment. The core feature of the DFT approach to sequence generation is illustrated here: The system operates stably in the face of such variable and unpredictable timing of each individual action. Moreover, the sensory signal sent to the condition of satisfaction system is obtained from the life camera image on the robot. The movement of the robot as well as intrinsic properties of the video system make this a noisy signal, but the bistable condition of satisfaction system stabilizes the decision that the action goal has been reached, leading to orderly transitions. Note also, how the duration of the transition itself varies, as reflected in the time interval during which the condition of satisfaction neuron is "on". This reflects the speed of the transition in the ordinal stack, which depends on the metric distance between the successive peaks, the strengths of the learned patterns, and on fluctuations in the neuronal dynamics.

3.6 Discussion

We have introduced a neurally grounded architecture for the learning and generation of behavioral sequences based on the framework of Dynamic Field Theory. To demonstrate the system's capability to tolerate variable durations of each action within a sequence, we implemented the architecture on an autonomous robot that searches for colored objects in a learned order of colors. The ordinal position of an action is encoded along a stack of neuronal activation fields, each of which expands feature dimensions needed to specify actions. This feature dimension thus represents the "contents" at each step, a form of positional encoding in the classification of (Henson & Burgess, 1997). While we have implemented a single feature dimension, color, in our robotic example, these fields may link to rich representations that bind multiple different feature maps into perceptual objects (Johnson *et al.*, 2008). By steering the perceptual system, this representation makes it possible to generate actions directed at objects in the world as illustrated in our robotic demonstration. How I overcome the limitation of this model – the homogeneity of actions involved in a sequence is presented in Chapters 5 and 6.

3. THE DFT SEQUENCE GENERATION ARCHITECTURE. A ONE-DIMENSIONAL MODEL.

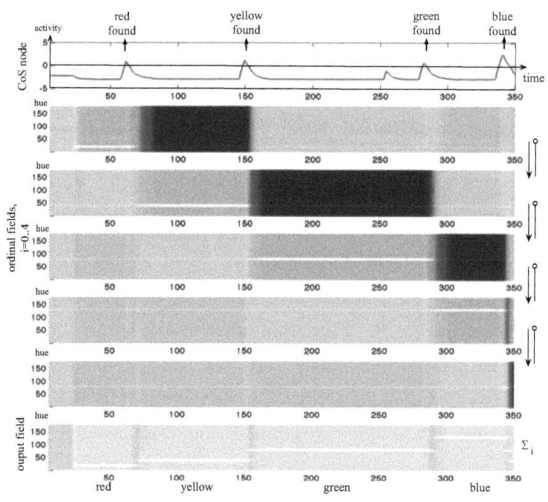

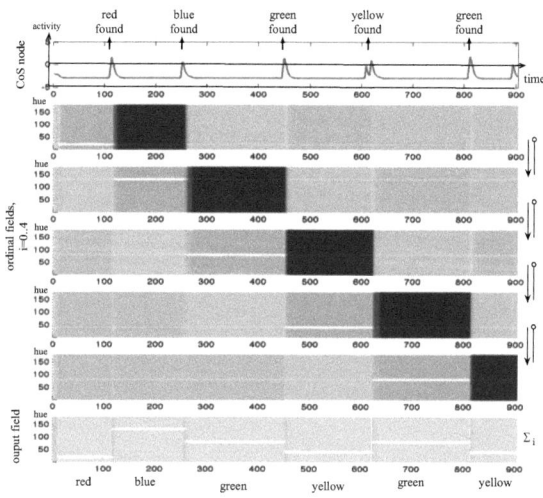

Figure 3.9: Time-courses of activity of the condition of satisfaction node, the ordinal fields, and the output field during production of the sequence "red-yellow-green-blue" (top) and the sequence "red-blue-green-yellow-green-yellow" (bottom).

4

Application of the DFT sequence generation mechanism to model turn taking

4.1 Embodied communication and turn taking: motivation for this application

The model of Chapter 3 provides a solution for the problem of how to control the timing of the initiation of an action by a sensed event that triggers an instability in the dynamics of the architecture. This capacity is important not only in autonomous robots, but in many other areas, in which a system is making decisions interspersed with processing. The field of embodied communication offers a relevant scenario: Humans excel at organizing their communicative behavior, e.g. they both produce and comprehend gestures, speech, they coordinate speech with gesture, and take turns. While in much theoretical work the detailed unfolding of communication in time is ignored, it is critical in real human communication. This becomes evident when you talk over a time-delayed transatlantic line, or even just in how much more cognitive load a phone conversation imposed compared to a real physical conversation. This is also a problem in technology: artificial dialogue systems are particularly limited with respect to flexible, well-timed interaction with human user (Becker *et al.*, 2004).

Here, we take an exemplary problem from the field of embodied communication – turn taking. In turn taking, the sequential switching of utterances between the two

4. APPLICATION OF THE DFT SEQUENCE GENERATION MECHANISM TO MODEL TURN TAKING

dialogue partners is at the core of a flexible and dynamic unfolding of a dialogue. Generating the amazingly smooth and fast transitions between turns requires that inherently time-variant events, the individual turns, be coordinated with good temporal precision. Some authors have postulated that turns are governed by oscillators, so that their durations are multiples of a base unit (Wilson & Wilson, 2005). This does not seem to capture the full level of temporal flexibility of embodied communication as reflected, in fact, by the distributions of silent intervals at turn switches observed experimentally by these very authors (Figure 4.3, left). These distributions have a lot of counts at surprisingly short times (of the order of 100 milliseconds!) but are still broad relative to their means. Another view is that turn taking requires anticipation, that is, predicting when a chance to switch may occur (Thórisson, 2002).

The turn taking model presented in this Chapter is primarily a metaphor for how the timing properties of turn taking may arise. The model does not have an explicit mechanism for anticipation, but turn switching is based on receiving a graded signal from the communication partner, which may, in effect, generate anticipation. The model does not address the rich inner structure of a dialog, nor the multi-modality of its embodiment, but demonstrates the important principles that may underly utterances exchange between partners. The model can provide a perspective for how the framework of dynamical systems thinking may help understand the autonomous, graded, and real-time structure of embodied communication.

In the next Section, I sketch the conceptual structure of the model. The mathematical equations are listed in Section 4.3, results from numerical simulations of the model are presented in Section 4.4. The results and impact of the model are discussed in Section 4.5.

4.2 The DFT model of turn taking

Consider two partners, "A" and "B", communicating with each other. Each actor can be modeled as an agent generating sequences of utterances. Each agent is thus a sequence generating system in my model. The neural fields of the model are defined over a metric dimension, x, that represents a feature value that characterizes each communicative act, utterance, or turn (Figure 4.1). As a simple example of this metric dimension, I use the planned duration of each communicative act, thus at the same time demonstrating how planned duration can be coordinated with turn-taking cues perceived by the speaker.

In this picture, there are two contributions to the contents of the items in a sequence. One represents a prior plan of a series of communicative actions. This "sketch of the conversation", or a number of arguments, that each partner has in mind before the communicative act, is modeled by localized inputs, or preshapes, associated with each step in the sequence. The other contribution is a reactive component. This is the influence of the current utterance on the listener's inner state and it is represented by localized input generated at each turn by the action of the other partner. In the simple example presented here, this is accomplished through a randomly generated mapping from the feature value of that other partner's action onto a feature value of the present partner's sequence plan. This "world model" can be learned or acquired in a developmental process in a more thorough consideration of embodied turn taking. Here I concentrate on the sequence generation aspect alone.

When the communicator "A" begins the planned sequence, a peak in the first ordinal field is located over the location of the corresponding preshape (left panel of Figure 4.1). This leads to a matching peak at location, x_{p1}, in that actor's motor field (bottom of the stack). Communication partner "B" might be at a different stage in its planned sequence of arguments. This communicator may have no peak at all (no communicative intention). More typically, however, this partner may have some communicative intention represented by a peak at a particular location in a particular ordinal position of its communicative plan. In the figure, partner "B" is at the second ordinal position in an ongoing sequence having generated a peak in the second ordinal field together with the matching peak at location, x_{p2}, in the motor field. Thus, communicator "B" faces the action of "A" already with some communicative intention, a prepared communicative act. Because of the mutually inhibiting actions systems (bottom of the Figure 4.1), partner "B" does not produce an utterance, its internal state stays stable, the perturbation from partner "A" is not sufficient to change the planned utterance of partner "B" (yet). When partner "A"'s utterance is about to end, a turn-taking mechanism, modeled by two coupled oscillators, competitive nodes, and condition of satisfaction nodes, comes into play: the action system of partner "B" can be activated now, while partner "A" proceeds in its sequence of planned utterances.

4. APPLICATION OF THE DFT SEQUENCE GENERATION MECHANISM TO MODEL TURN TAKING

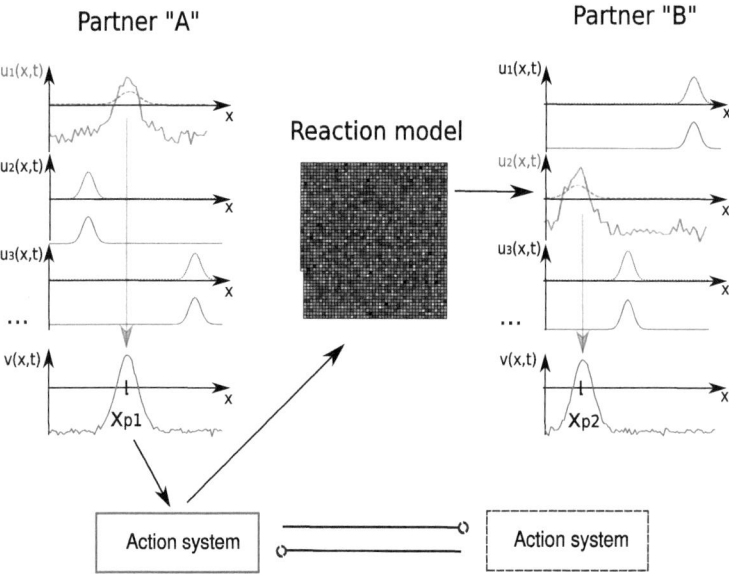

Figure 4.1: The DFT turn taking architecture consists of a DFT sequence generation model for each of the two communication partners, "A" and "B" (stacks of fields on the left and on the right). Action system that generates communicative acts is modeled by a neuronal oscillator that deactivates itself after a single oscillation. The action systems of the two partners compete, so that only one is active at a time. The reaction model provides localized input into the sequence generation system of one partner ("B" here) based on the current action of the other partner ("A" here), modeling the influence of the current utterance on the state of the listener.

In the following Section, I describe the turn taking model in terms of neural field equations. The functioning of the model is exemplified in Section 4.4.

4.3 Mathematical description of the turn taking model

4.3.1 The sequence generation mechanism.

The dynamic fields, $u_i^A(x,t)$, represent the ordinal position, $i = 1, \ldots, N$, of items in the sequence of a communication partner, A, along the feature dimension, x, and evolve in time, t, according to equation Eq. (4.1).

$$\begin{aligned}
\tau_o \, \dot{u}_i^A(x,t) &= -u_i^A(x,t) + h_o^A + \int f(u_i^A(x',t)) w_{oo}(x,x') dx' \\
&+ C_+ F_{\text{env}} \int f(v^A(x',t)) f(u_{i-1}^A(x',t)) dx' - C_- \int f(u_{i+1}^A(x',t)) dx' + P_i^A(x,t) \\
&+ \int \text{RM}(x,x') f(v^B(x',t)) dx'. \quad (4.1)
\end{aligned}$$

The analogous equation for the communication partner, B, is obtained by switching upper indices A and B. Here, τ_o is the time constant of the field dynamics, h_o^A is a constant resting level of the field, $f(u) = 1/(1 + e^{-\beta(u-u_o)})$ is a sigmoidal function, where β is a parameter and u_0 is a threshold. $w_{oo}(x,x') = -c_{\text{inh}} + c_{\text{exc}} e^{-(x-x')^2/2\sigma^2}$ is an interaction kernel, where parameters c_{inh} and c_{exc} are strengths of inhibitory and excitatory interactions. $C_+ F_{\text{env}}$ models the sensori-motor feedback about accomplishment of the current action, defined below; C_- is the strength of backward inhibition along the stack; $P_i(x,t)$ is a localized pre-activation of the ordinal field number i encoding what is represented or planned at that ordinal position in the sequence; $\text{RM}(x,x')$: is an $N \times N$ matrix ('reaction model') that models communication by associating an output of partner B with input to partner A's ordinal stack. Here it is a fixed pseudo-random matrix.

The time-dynamics of activity in the motor field, $v^A(x,t)$, of partner, A, is governed by equation, Eq. (4.2).

$$\begin{aligned}
\tau_M \dot{v}^A(x,t) &= -v^A(x,t) + h_m^A + \int f(v^A(x',t)) w_{mm}(x,x') dx' \\
&+ \Sigma_{i=0}^N \left[\int f(u_i^A(x',t)) w_{mo}(x,x') dx' + C_+ \int f(u_i^A(x',t)) dx' \right] \quad (4.2)
\end{aligned}$$

4. APPLICATION OF THE DFT SEQUENCE GENERATION MECHANISM TO MODEL TURN TAKING

Again, the analogous equation applies to the communication partner, B. The following terms are added in this equation: h_m^A is the resting level, $w_{mm}(x, x')$ is an interaction kernel analogous to $w_{oo}(x, x')$, $w_{mo}(x, x')$ is a Gaussian-shaped projection kernel from the ordinal position stack to the motor field.

4.3.2 The turn taking mechanism.

The sensori-motor system that generates a communicative act is described by two action variables, x_A and y_A. In the present model, the only significance of these action variables is to signal that an action is ongoing. The dynamics of these variables is the Hopf normal form, Eq. (4.3).

$$\begin{aligned}\tau_h \dot{x}_A &= g_A \left[\gamma(\mu - x_A^2 - y_A^2)x_A - \omega_A y_A\right] - (1 - g_A)x_A \\ \tau_h \dot{y}_A &= g_A \left[\gamma(\mu - x_A^2 - y_A^2)y_A + \omega_A x_A\right] - (1 - g_A)(y_A - 1)\end{aligned} \quad (4.3)$$

This dynamics generates a stable limit cycle for partner A. The analogous equation applies to partner B. The notion is as follows: τ_h is time scale of the oscillator dynamics, γ is a parameter determining the relaxation rate of the limit cycle, μ is a parameter determining the amplitude ($= 2\sqrt{\mu}$) of the limit cycle, g_A (and, analogously, g_B) are dynamic neuronal activation variables described below that turn on ($g_A = 1$) and off ($g_A = 0$) the stable limit cycle solution. When these variables are in an "off" state ($1 - g_A = 1$), the dynamics has a fixed point at ($x_A = 0, y_A = 1$). ω_A determines the frequency of the limit cycle and is determined from the motor field by equation, Eq. (4.4).

$$\omega_A = C \int x f(v^A(x,t)) dx, \quad (4.4)$$

where the constant, C, is a normalization factor. This is an estimate of the location of a peak within field, $v(x,t)$.

The neuronal activation variables, g_A and g_B, evolve according to a competitive dynamics that is built from two normal forms of the degenerate pitchfork bifurcation, coupled competitively, Eq. (4.6).

$$\tau_g \dot{g}_A = \alpha_A g_A - |\alpha_A|g_A^3 - g_B^2 g_A + \xi_A \quad (4.5)$$

$$\tau_g \dot{g}_B = \alpha_B g_B - |\alpha_B|g_B^3 - g_A^2 g_B + \xi_B \quad (4.6)$$

The coupling makes that only one neuron may be activated at a time, e.g., $g_A = 1$ and $g_B = 0$ (see (Schöner & Dose, 1992) for a mathematical analysis). The factors, α_A and α_B, determine which of the two neurons is activated. These factors are designed to be positive (enabling the associated neuron to be turned on) only if there is a peak in the motor field of the corresponding communication partner. They become very small when the associated oscillator is near one end of its limit cycle (e.g., $x_A \approx -\sqrt{\mu}$), generating a tendency for the associated oscillator to be turned off (see an empirical formula for partner "A" in Eq. 4.7).

$$\alpha_A = \Big(-2f(2e^{-100(x_A+\sqrt{\mu})^2}) + 0.67 \Big) \cdot \Big(c \int f(v^A(x,t)) dx \Big). \tag{4.7}$$

Here, c is a normalizing factor, f is a sigmoidial non-linearity, other variables are the same as in Eq. 4.6 and Eq. 4.3.

These factors, α_A and α_B, also serve to generate the signal, F_{env}, that provides the "condition of satisfaction" to the stack of ordinal fields. For partner "A", the condition of satisfaction signal equals $cf\Big(2e^{-100(x_B+\sqrt{\mu})^2}\Big)$ under condition that g_B is close to 1.

4.4 Results

The model of turn taking is a dynamical sketch of the action system that each communicator uses to "act out" communicative intentions. These systems generate single "actions" with a well-defined duration by starting a limit cycle oscillator with a specific frequency. In fact, the frequency of the oscillator and thus the duration of its "action" is encoded along the metric dimension of the sequence generation systems. While other parameters of the utterances can be encoded in the same manner, different durations of communicative acts constitute the most interesting aspect with respect to the necessity to stabilize the representation of each planned utterance.

Each oscillator has two inner state variables. One of these variables rises from zero, reaches a maximum, and falls back to zero in each cycle of the oscillator. After a single cycle, the oscillator is turned off through the competitive dynamics and a dynamic variable monitoring the phase of the oscillator. This activation variable can be either in an "on" state or an "off" state. Input from the motor field and absence of competition from the other actor turns the activation variable "on". When the associated oscillator

4. APPLICATION OF THE DFT SEQUENCE GENERATION MECHANISM TO MODEL TURN TAKING

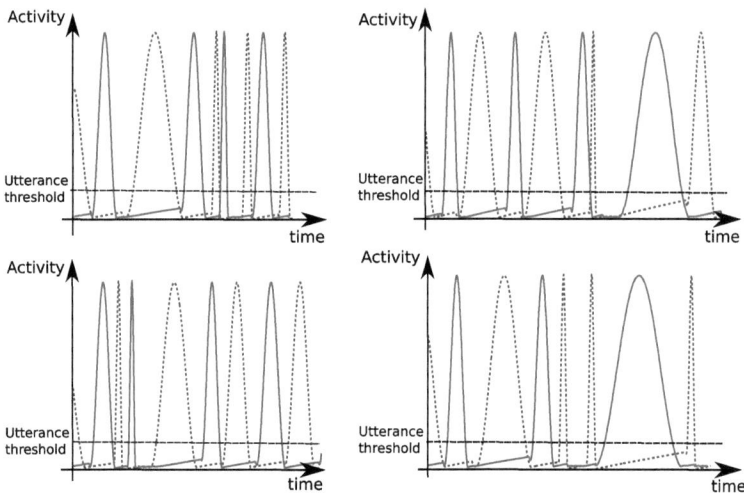

Figure 4.2: Activity of the oscillators that model action systems of the actors "A" (solid lines) and "B" (dotted lines) during four sample "dialogues" generated by the turn taking model. Positive activation (beyond the threshold marked with a solid line) corresponds to an "utterance". Periods of time when both oscillators are inactive correspond to silence intervals, the duration of which contributes to the histogram shown in Figure 4.3.

reaches the end of a cycle, the activation variable is switched to an "off" state through a dynamic instability.

The level of the activation variables controlling each oscillator is exchanged between the two communicators as a signal for how close they are to yielding the turn. A large level of activation of action system "A" inhibits action system "B" and vice versa.

Figure 4.2 illustrates simulations of this complete model by displaying the time courses of a state variable characterizing the respective oscillators of both actors. Episodes of oscillator activation are individual communicative actions. These episodes vary in duration as dictated by the frequency of the limit cycle oscillators encoded by the location of peaks in the neural fields of the sequence generation systems. An action system, whose state variable is below a threshold, is in the "off" state waiting its turn.

4.5 Discussion of the turn taking model

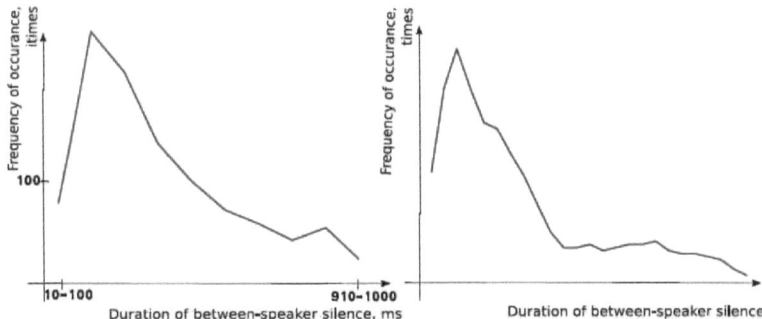

Figure 4.3: Left: a histogram of the durations of silence intervals between turns observed experimentally in a human conversation, data provided in (M. Wilson & Wilson, 2005). **Right**: A corresponding histogram generated by the DFT turn taking model. As the time scale in the model is arbitrary, only the overall shapes of the distributions are compared.

When the state variables of both action systems are below threshold, then both action systems are in the "off" state and a period of silence between turns is observed. The amount of silence at each turn switch varies because the state variables change at variable rates due to the variable durations of actions. Moreover, turn switches involve instabilities in the dynamics of the action systems. This gives noise a sizeable influence on the exact time at which a transition is realized. As a result, the histogram of the durations of silent intervals obtained from an ensemble of simulation runs shown in Figure 4.3 is quite broad, although centered on a most frequent interval. This characteristic of a small mode and long tail matches the shape of the distributions obtained from human data shown in the same figure (left on the Figure 4.3).

4.5 Discussion of the turn taking model

This simple mathematical model of turn taking is a dynamical system's "finger-print" of the time structure of embodied communication. It offers a means to model this time structure within the neuronally based theoretical framework of the dynamical systems approach to embodied cognition. Applying the DFT sequence generating system to model the sequence of utterances that constitute a dialogue makes it possible to *au-*

4. APPLICATION OF THE DFT SEQUENCE GENERATION MECHANISM TO MODEL TURN TAKING

tonomously generate the time-course of a dialogue. The utterances here are not paced by a rigid input-compute-output cycle, but are open to sensory input at any time. Because the two systems are almost always in a *stable* state, such continuous coupling does not prevent the systems from performing their assigned function. The systems are sensitive to input only while near an instability, which leads to a switch of turn.

These instabilities start at the level of the action system, which is turned "on" and "off" in response to graded signals received from an internal dynamical timer as well as from the communication partner. The instabilities amplify small graded changes of signals into macroscopic changes. This demonstrates how dynamical systems can make sense of and depend on *graded* variables while at the same time being able to make categorical decisions and to discard graded information as they set in motion a new action step.

This simple account models the action systems as self-controlled, one-shot oscillators that stand in for the much richer action systems engaged when people communicate. These include the speech articulatory system, systems controlling the prosody of their utterances, controlling gesturing, controlling facial expression and body posture. All these systems may have graded components, from which communication patterns may derive signals for how close the actor is to yielding the turn. That we model such complex systems as stable limit cycle oscillators is a deliberate scientific move. The coupling among stable oscillators is argued to be the basis for coordination of timed actions (Schöner, 2002; Schöner & Kelso, 1988). Thus, it is easy to imagine how the multiple actions involved in generating communicative acts may be coordinated with each other as well as across two communication patterns through couplings of the kinds modeled in simplified form here. This metaphor may provide an avenue toward an account for the remarkable temporal regularity and order observed in embodied communication (Streek, 1993).

The meaning transmitted in and expressed through communicative acts has been modeled only minimally here. In DFT, meaning is encoded through continuous metric dimensions. The location of peaks along such dimensions signifies specific instances of such metric information. In principle, coupled networks of dynamic fields may generate distributed representations of perceptual objects (see (Johnson *et al.*, 2006) for an example). The multiple modalities and perceptual dimensions that are relevant to embodied communication could easily be cast in dynamic field terms. Within the

4.5 Discussion of the turn taking model

toy model presented here, the field dimension represents the duration of a planned communicative act and in that respect acts merely as a place holder for more substantive communicative meaning.

The field dynamics provides a framework for integrating different sources of specification of meaning. For instance, a memory trace of previously activated or stimulated patterns of activation may bias the ordinal fields to particular locations. On the other hand, current input from the other actor's communicative act may be overlaid and fused with such a prior plan. This may lead to the selection of activation patterns that in some way match the received message. The metrics of patterns of activation within continuous activation fields forms the basis for determining such matches. This metrics can be exploited by conceiving of structured forward mappings, like those of connectionist networks, to replace the random world matrix of our toy model.

Thus, much of the machinery needed to enrich the processing exists within connectionist and dynamical systems approaches. The challenge will be to translate insights obtained from substantive models of the information processing involved in embodied cognition (Thórisson, 2002) into dynamic terms. What is needed for this to happen is that aspects of embodied communication that provide entry points into dynamical systems thinking are identified and collaboratively explored in both theory and experiment.

This work demonstrates how the DFT sequence generating model can be applied to build generative models of cognitive processes within or between embodied agents. The behavior of the model can be compared to human performance, and can be used to understand the processes involved in cognition.

4. APPLICATION OF THE DFT SEQUENCE GENERATION MECHANISM TO MODEL TURN TAKING

5

Ordinal dynamics based on discrete activation variables

The sequence generation model presented in Chapter 3 is based on an ordered stack of identical neural fields. In this model, the temporal component of a sequence spans a dimension along with a "spatial" dimension of the motor or perceptual parameter that characterizes actions. The temporal dimension is sampled by the set of ordinal neural fields. In a task with a richer behavioral repertoire, the dimensionality of the full action system of the agent might be very high: action fields would be spanned over spaces of many dimensions, or characteristic variables. Such high-dimensional neural fields, however, are neither computationally trackable (there would simply not be enough neurons in the cortex to sample such high-dimensional metric spaces), nor conceptually justified. Indeed, actions and percepts may originate from different "modalities", e.g. they may be coupled to different effectors and sensors, and not interact with each other according to the neural field dynamics. Therefore, we may separate the behavioral dimensions that underly different kinds of actions, e.g. we may separate color search from arm movement.

In this picture, several lower-dimensional dynamic neural fields represent actions of different modalities. In order to implement sequences of such actions, the mechanism of preshaping the neural fields defined over the same dimension must be replaced by some more flexible dynamic mechanism that enables pointing to appropriate feature spaces. This mechanism may be implemented in a set or discrete dynamical nodes, that "stand for" actions in the particular ordinal positions. The content of the actions

5. ORDINAL DYNAMICS BASED ON DISCRETE ACTIVATION VARIABLES

that is encoded in dynamic neural fields is specified by weighted connections from the discrete nodes to the fields. The discrete nodes are not embedded in a single metric space. In this respect, they represent actions symbolically, they create instances of categories of action, namely those that occur at a certain ordinal position in a sequence. The condition of satisfaction system must be extended in a case of such multimodal actions: it now must explicitly represent which action modality and which value of the characteristic variable of this modality it is associated with. Therefore, a neural field replaces the condition of satisfaction node of the former architecture.

Figure 5.1 compares two DFT sequence generation architectures visualizing the core ordinal dynamics. The left part of the figure represents the sequence generation model of Chapter 3 with dynamic neural fields encoding actions' metric parameters along with the ordinal information. The right part of the figure represents a modified architecture, in which the ordinal information is represented by a set of discrete nodes that project their activity onto a dynamic neural field that represents actions. The condition of satisfaction (CoS) signal is generated by a CoS dynamic neural field in the latter model, as described in the following Sections of this Chapter.

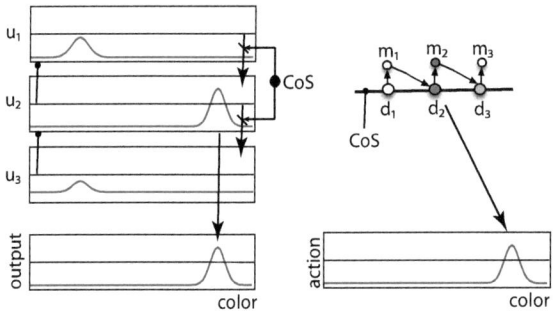

Figure 5.1: The DFT sequence generation models with representation of ordinal information within continuous neural fields (left) and within a discrete set of ordinal nodes (right), an overview.

In the following, I introduce the DFT sequence generation architecture with discretely represented ordinal information. Compared to the previous model, in this architecture: (1) neural field representations of different actions are combined with a neuronal dynamics of discrete nodes that represent ordinal positions in a sequence of actions; and (2) the condition of satisfaction system is extended to represent a range of perceptual parameters that characterize final states of the actions. To draw a better comparison to the DFT sequence generating model of Chapter 3, the same color-search scenario is used here in order to demonstrate that the modified architecture is also capable of embodied sequence generation and satisfies the constraints I described in the Introduction. This Chapter, thus, does not yet exploit the potential for heterogeneity. That will happen in Chapter 6.

5.1 The extended DFT sequence generation architecture

An overview of the DFT sequence generation architecture is given in Figure 5.2.

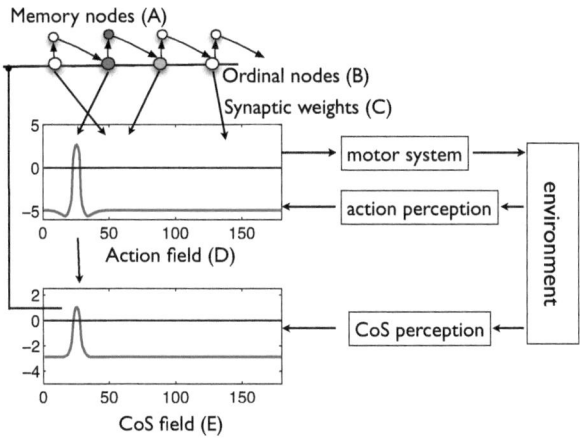

Figure 5.2: Functional modules of the DFT sequence generation architecture.

First, we step through the main functional components of the architecture. A set of discrete neural-dynamic nodes represents ordinal positions within a sequence at which the system is at any point in time (Figure 5.2B). This set of *ordinal nodes* is inspired by neurophysiological evidence for a neuronal representation of serial position within a sequence of actions. Neurons responsive to ordinal information have been found in

5. ORDINAL DYNAMICS BASED ON DISCRETE ACTIVATION VARIABLES

neostriatal cortex (Aldridge & Berridge, 1998), in motor cortex (Carpenter *et al.*, 1999), and in the supplementary motor area (Clower & Garrett, 1998). The ordinal nodes in our architecture are bistable activation dynamics that share inhibition so that only one ordinal node can be active at a time. Asymmetrical excitatory connections between the nodes control the direction in which activation spreads from the beginning to the end of a sequence. These directed connections are mediated by *memory nodes* (Figure 5.2A) that stay active during a transition phase between two successive actions.

When an ordinal node is activated, its projection onto an action field becomes effective. One or multiple *action fields* (Figure 5.2D) span the space of actions that are possible at any given stage of the sequence. Which action is associated with a given stage is specified by the *synaptic connections* (Figure 5.2C) from the corresponding ordinal node to the action field. During sequence learning, those connections are strengthened that project onto the field sites that are activated by the demonstrated action. The set of synaptic weights thus constitutes the memory for a graded representation of the action associated with a particular ordinal position. As an ordinal neuron becomes active, it induces a self-stabilized peak of activation at the location to which the ordinal neuron projects. Such a peak in an action field then elicits motor behavior. The *motor system* is a dynamics that is controlled by the movement parameter values, over which the peak in the action field is centered. This dynamics couples to the effector system of the agent. The action field may receive perceptual input from the environment while a particular action is being executed or observed. Through this perceptual channel, peaks of activation may be induced in the action field during sequence learning in response to a demonstration of a target action. During sequence production, the perceptual input is not sufficient to induce an activity peak, which is accomplished by lowering the resting level. The *action-perception* dynamics controls how this perceptual input is extracted from the sensory surface.

To solve the problem of destabilizing the stable state that specifies the current action we explicitly represent the fact that this action has achieved its goal. This is inspired by the concept of a condition of satisfaction in Searle's analysis of intentional acts (Searle, 1983). The third component of the architecture is, therefore, a *condition of satisfaction (CoS) field* defined over a relevant perceptual dimension specific to the space of possible actions (Figure 5.2E). The CoS field receives a localized input from the active action field. This input pre-activates those locations in the CoS field that respond to perceptual input characteristic for the state when the currently specified action is finished. This makes the CoS field more sensitive to this particular sensory input. The sensory input is obtained from the *CoS perception* dynamics that is structured to pick up signals relevant to the potential termination of actions. Where that sensory input overlaps along the field dimension with the localized input specified by the action field, the detection threshold is reached and a self-stabilized peak is formed in a detection instability. That peak inhibits the ordinal set and triggers a cascade of instabilities in the sub-systems of the architecture, the end-result of which is that the next ordinal node is activated and the system proceeds to executing the next action.

5.2 Mathematical structure of the DFT sequence generation architecture with discrete ordinal nodes

5.2.1 Ordinal nodes

The ordinal system consists of ordinal nodes with activities, d_i, and associated memory nodes with activities, d_i^m, where the index, i, counts the actions. The activity of these nodes evolves in time according to a set of coupled dynamic equations:

$$\tau \dot{d}_i(t) = -d_i(t) + h_d + c_0 f(d_i(t)) - c_1 \sum_{i' \neq i} f(d_{i'}(t)) + c_2 f(d_{i-1}^m(t))$$
$$- c_3 f(d_i^m(t)) - I_C(t) \qquad (5.1)$$

$$\tau \dot{d}_i^m(t) = -d_i^m(t) + h_m + c_4 f(d_i^m(t)) - c_5 \sum_{i' \neq i} f(d_{i'}(t)) + c_6 f(d_i(t)) \qquad (5.2)$$

To understand the role of the different terms, it is useful to examine the state of this system at three critical moments in time during a transition illustrated in Figure 5.3 as well as the associated time courses and dynamics shown in Figure 5.4. The first three terms in Eq. (5.1) are a dynamic stabilization factor $-d_i(t)$, a negative resting level, $h_d < 0$, and a self-excitatory term of strength, c_0. Herein, $f(\cdot)$ is the sigmoidal non-linearity of Eq. (2.3). Together, these terms establish a bistable dynamics of each ordinal node which can be either in an "off" attractor ($d_i < 0$) or an "on" attractor ($d_i > 0$) (see top row of Figure 5.4). Inhibition among all ordinal nodes of strength, c_1, allows only one node to be "on" at any time. The ordinal node at the ith step is excited with strength, c_2, when the memory node of the previous, $(i-1)^{\text{th}}$, step is "on", but is inhibited with strength, c_3, when the memory node at the same, i^{th}, step

5. ORDINAL DYNAMICS BASED ON DISCRETE ACTIVATION VARIABLES

is "on". The first of these inputs drives activation of the next step, the second helps destabilize the predecessor of the new action. Loss of stability of the activated predecessor state requires in addition an inhibitory input, $I_C(t) = c_{CoS} \cdot \int_y f(U_C(y,t))dy$, from the condition of satisfaction field, $U_C(y,t)$, Eq. (5.5).

Memory nodes have the same basic bi-stability between "on" and "off" states established by the first three terms of Eq. (5.2). The memory modes are inhibited with strength, c_5, by any activity in the ordinal layer, except for activity at the corresponding ordinal node, which excites with strength, c_6. The strength, c_4, of self-excitation is chosen such that the "on" state is sustained after this excitatory input has been removed, so that the memory node stays "on" after the sequence has advanced to the next step.

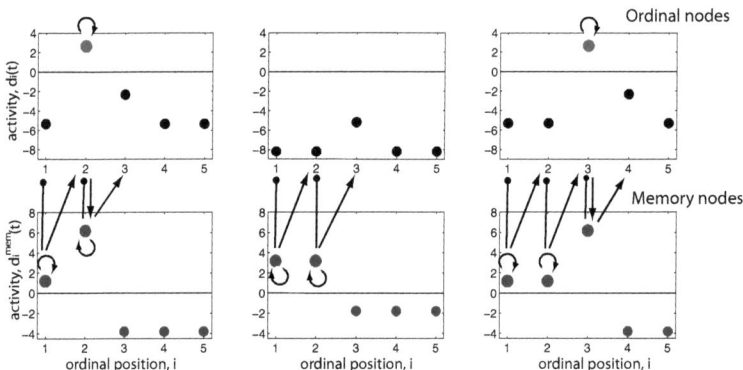

Figure 5.3: The activation levels of the five ordinal (top) and memory (bottom) nodes at three time slices during a transition from the second to the third action in a sequence. **First column:** the state of the ordinal system while the second action in a sequence is produced, the second ordinal node is activated and specifies the ordinal position. **Second column:** the transition phase, when a condition of satisfaction signal inhibits the ordinal pool. The third ordinal node is less negative, because it receives input from the second memory node, but is not inhibited by the third memory node. **Third column:** after the condition of satisfaction signal ceases, the least negative third ordinal node reaches the threshold for a detection instability and gets activated. Arrows show active excitatory connections, lines ending with a filled circle show active inhibitory connections.

5.2 Mathematical structure of the DFT sequence generation architecture with discrete ordinal nodes

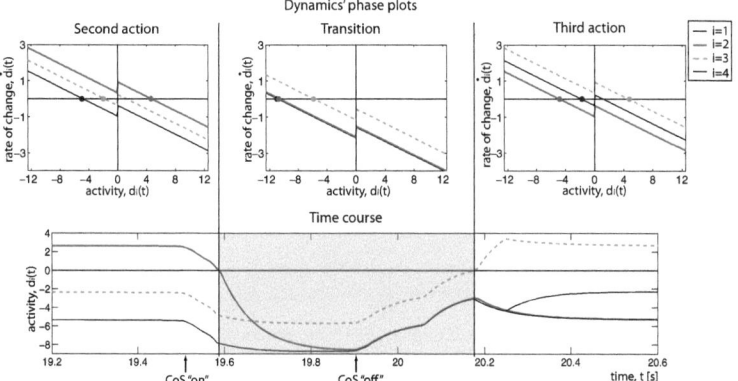

Figure 5.4: Top: Visualization of the ordinal dynamics at the same three time slices during a transition, as in the Figure 5.3. Intersections with abscissa, marked by circles, are stable fixed points corresponding to "off" (negative half-space), or "on" (positive half-space) states. **Bottom:** The time course of ordinal nodes' activations during a sequential transition. The points in time when the condition of satisfaction signal enters the ordinal set (CoS "on") and ceases (CoS "off") are marked by the arrows. The shaded region corresponds to the transition phase between the two actions.

5. ORDINAL DYNAMICS BASED ON DISCRETE ACTIVATION VARIABLES

5.2.2 Action fields and synaptic projections

Ordinal nodes specify an action by projecting onto an action field, a dynamic representation of a continuum of possible actions spanned by an appropriate continuous parameter, x. The dynamics of the action field,

$$\tau \dot{U}_A(x,t) = -U_A(x,t) + h_A + \int f(U_A(x',t))w(x-x')dx' + c_{ord}\sum_{i=1}^{N_d} f(d_i(t))M_i(x,t)$$
$$+ c_{vis}I_{vis}$$
(5.3)

is set up by the first three terms, which define the generic neural field dynamics of Eq. (2.1). The projection from the ordinal nodes, $f(d_i(t))$, is mediated by modifiable weights, $M_i(x,t)$, that specify the connection strength between the ith ordinal node and the site, x, of the action field. The sum is over all N_d ordinal nodes, c_{ord} is a constant controlling strength of the ordinal contribution. Last term is a perceptual input which strength is controlled by the factor, c_{vis}, that is set to larger values during sequence learning.

If the non-zero connection weights, $M_i(x,t)$, are localized around a particular action characterized by the parameter, x_0, then activation of the associated ordinal node ($f(d_i(t)) > 0$) pushes sites around x_0 above threshold and a self-stabilized peak is induced in the action field via a detection instability. That peak is a stable neural representation of the action linked to the ith step of the sequence. Projections from the action field onto the motor surface steer the system so that it executes an appropriate physical action. An exemplary implementation of such motor behavior is given in Section 5.3 "Robotic implementation". Figure 5.5 illustrates how the ordinal set determines the location of a peak in the action fields by comparing two moments in time corresponding to the second (left) and the third (right) step.

The weights of the neural connections, $M_i(x,t)$, are modified during learning according to a Hebbian-like rule, Eq. (6.2):

$$\tau_l \dot{M}_i(x,t) = \Big(-M_i(x,t) + f(U_A(x,t))\Big) \cdot f(d_i(t))$$
(5.4)

Thus, when an ordinal node is active ($f(d_i(t)) > 0$), the weights are strengthened to those sites of the action field at which positive activation ($f(U_A(x,t)) > 0$) represents

5.2 Mathematical structure of the DFT sequence generation architecture with discrete ordinal nodes

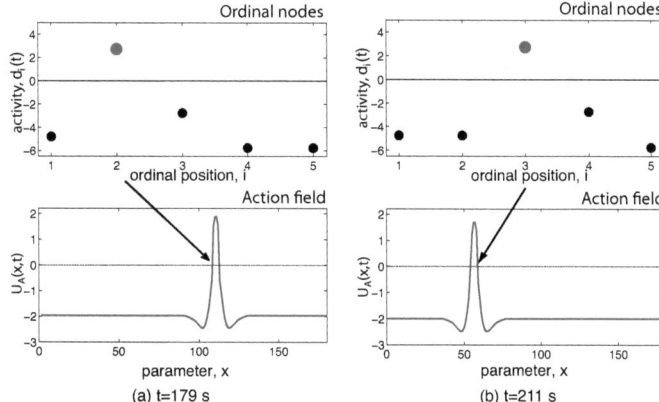

Figure 5.5: Projections from the ordinal nodes onto the dynamic field representing a neural parameter of action. (a) At time $t = 179s$, the second ordinal node, $d_2(t)$, is active and induces the representation of the associated action in the action field, $U_A(x,t)$. (b) At time $t = 211s$, the third ordinal node is active and induces a different peak in the action field that represents the third action. Arrows indicate centers of the learned localized projections from the ordinal nodes to the action field.

the currently demonstrated action. Below we will show how sensory input during learning induces localized activation peaks in the action field. Converging to the output of the action field on the time scale of the learning dynamics, τ_l, the weight matrix, $M_i(x,t)$, assumes the localized form in which a single action is associated with each ordinal node.

5.2.3 Condition of satisfaction field

For each action in a sequence, its termination criterion, or *condition of satisfaction* (CoS) is represented neuronally in a dynamic field, $U_C(y,t)$, defined over a dimension, y, that captures the sensory states identifying the terminal criterion of an action and need not be identical to the characteristic dimension, x, of the action field, Eq. (5.3).

5. ORDINAL DYNAMICS BASED ON DISCRETE ACTIVATION VARIABLES

The CoS neural field evolves according to a dynamical equation Eq. (5.5):

$$\tau \dot{U}_C(y,t) = - U_C(y,t) + h_C + \int f(U_C(y',t))w(y-y')dy' \\ + c_A \int T(x,y)f(U_A(x,t))dx + c_{vis}^C I_{vis}(y,t). \quad (5.5)$$

The first three terms here set up the usual Amari field dynamics with time constant, τ, resting level, h_C, and interaction kernel, $w(y-y')$. Positive activation in the action field $(f(U_A(x,t)) > 0)$ propagates to the CoS field through the synaptic weights matrix $T(x,y)$ that defines the mapping between the dimensions characterizing actions, x, and their terminal states, y. Sensory input, $I_{vis}(y,t)$, of strength, c_{vis}^C, induces peaks in the CoS field when it overlaps in dimension, y, with input from the action field. c_A is a constant.

The function of the CoS field is illustrated in Figure 5.6. A peak of activation in the action field (panel A) represents a particular action and provides localized input to the CoS field (B), making it sensitive to sensory input (C) at matching sites. When the terminal state of the action is detected by the sensory system (F), sensory input to the CoS field overlaps with the preshaping input from the action field (E). The integrated inputs are sufficient to surpass the activation threshold of the CoS field and a self-stabilized peak arises. The positive activation in the CoS field signals the successful accomplishment of the on-going action. This neural signal is propagated to the ordinal set, where it uniformly inhibits all ordinal nodes and suppresses any positive activity in the ordinal systems. The signal triggers an instability in the dynamics of the system, which is described in the following section.

5.2.4 The sequential transition as a cascade of instabilities

The concurrent evolution of the ordinal system, the action field, and the CoS field linked to the environment through motor behavior and sensory signals leads to the sequential transition from one action to the next in a cascade of instabilities in the dynamical sub-systems of the architecture which are illustrated in Figure 5.7.

We look at the system when the second action in the sequence has been activated and controls the motor behavior. The cascade of instabilities begins when the terminal condition of that action is picked up by the sensors. This leads to positive activation

5.2 Mathematical structure of the DFT sequence generation architecture with discrete ordinal nodes

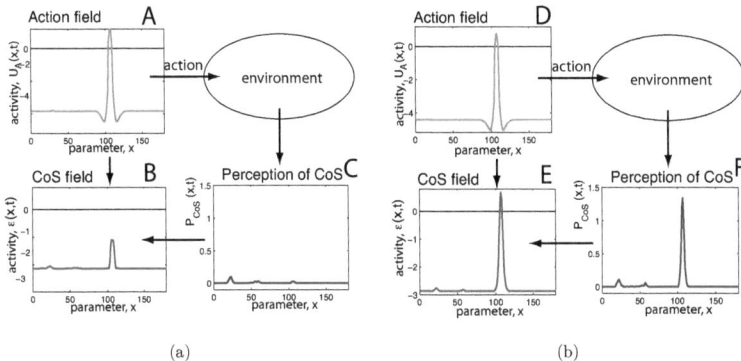

Figure 5.6: Detection instability in the condition of satisfaction (CoS) field. **(a)** During the action production, the CoS field (B) is preshaped by the action field (A) to be sensitive to the desired end-state of an action, that can be detected by the perceptual module (C). **(b)** Beginning of the transition phase: When sensory input (F) matches the preshape from the action field (D), a peak in the CoS field (E) is induced and stabilized.

in the perceptual system, which feeds into the CoS field. There, the perceptual input matches the preactivation from the action field. The combined input is capable of pushing the CoS field through the detection instability, leading to a self-stabilized peak. The new CoS peak inhibits the ordinal system, pushing all nodes including the activated second one below threshold (a reverse detection instability). As a result, input is removed from the peak in the action field, pushing it beyond the reverse detection instability, so that this peak decays. Its decay removes critical input from the CoS field, triggering a reverse detection instability there, so that the CoS peak decays. This removes global inhibition from the ordinal system. Within the ordinal system, the third node has a competitive advantage because it is boosted by the memory node of the second step. It goes through the threshold first, stabilizes its "on" state, and inhibits the other ordinal nodes. The projection of the third node onto the action field pushes that field through the detection instability at the location associated with the third node. The new peak in the action field impacts on the motor behavior of the system. The system has the equivalent dynamic state as at the beginning of the cascade, but now with the third rather than second sequential action in execution.

5. ORDINAL DYNAMICS BASED ON DISCRETE ACTIVATION VARIABLES

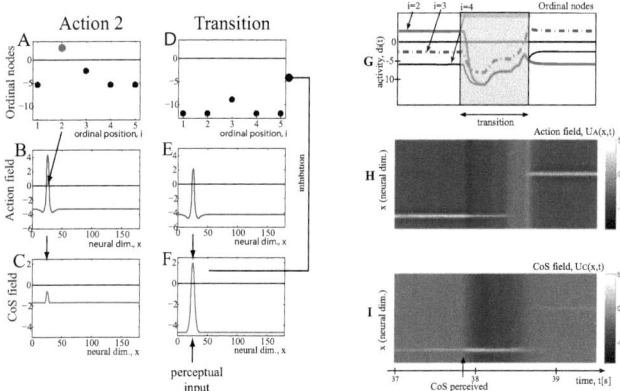

Figure 5.7: A transition between two actions in a sequence. **Left:** The activation of the ordinal nodes, the action field, and the condition of satisfaction (CoS) field during the second action. The second ordinal node (A) projects onto a location, $x \approx 30$, of the action field (B), and induces a self-stabilized peak there that represents the on-going action. The CoS field (C) is pre-activated at the matching location. **Middle:** The same variables after a sensory signal has been perceived that matches the terminal condition of the second action. This induces a peak of activation in the CoS field (F), which inhibits the ordinal pool (D). The peak in the action field decays (E). **Right:** Time courses of the ordinal nodes (G), the action field (H), and the CoS field (I) during a sequential transition. The first event in the cascade of instabilities of the transition is the build-up of a new peak in the CoS field (detection instability). The peak inhibits all ordinal nodes below threshold and leads to the decay of the peak in the action field (reverse detection instability). Next, the peak in the CoS field decays (reverse detection instability), which leads to the release of the ordinal system from inhibition and the activation of the third ordinal node. Finally, a new peak is formed in the action field (detection instability), which represents the third action.

Note that each instability within a cascade takes place within one of the dynamic fields or discrete neural dynamics. In terms of the complete overall dynamics, a cascade is a single transient triggered by the detection instability in the CoS field. The robust temporal order of the instabilities within the cascade is established by the structure of the neuronal architecture.

The same cascade of instabilities is at work during sequence learning. During learning, the peak in the action field is generated by direct sensory input that originates from a demonstration of the action. The co-activation of an ordinal node and the action field drives the synaptic weights from the ordinal node to the action field. Learning of a step in the sequence is terminated by generating a sensory signal that feeds into the CoS field. This sensory signal can be delivered by an autonomous action segmentation mechanism. The signal triggers the same cascade of transitions illustrated above and makes the system ready to learn the next sequential action. An exemplary implementation of the motor and perceptual systems, and a demonstration of the components' integrated dynamics guiding real-world actions are presented next.

5.3 Robotic implementation

To demonstrate how the DFT sequence generation architecture can be embodied, that is linked to simple sensory and motor systems, we implement the architecture on an autonomous robot in a simple task setting, a sequential color-search task. Presenting differently colored objects to the robot's camera in a given order, the system is "taught" a sequence of colors (Figure 5.8a). Here, learning is not supervised in the traditional sense in that the sensory signal that drives learning is not an error signal that informs about what would have been the correct response. Instead, the user's supervision consists of demonstrating the sequence by creating environmental conditions conducive to generating relevant sensory inputs. The acquisition of these inputs is autonomous.

Controlled by a simple behavioral dynamics that comprises target acquisition and obstacle avoidance, the robot is capable of searching for an object of a given color within an arena, in which colored objects are distributed (Figure 5.8b). The robot does so in the order in which the colors were presented during learning. For instance, if taught the sequence "red-blue-green-red-yellow", the robot will first search for a red object, then for a blue object, then for a green object, and so on. Each time an object

5. ORDINAL DYNAMICS BASED ON DISCRETE ACTIVATION VARIABLES

of the currently requested color looms large on the robot's visual array, that object is considered "found" and the robot switches to the next color in the sequence.

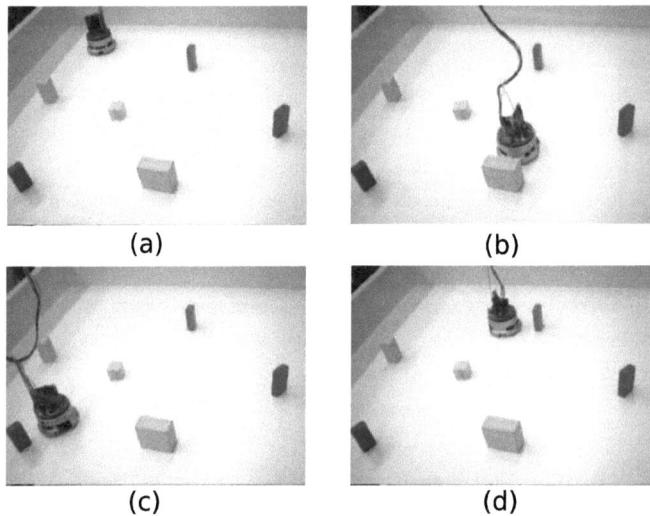

Figure 5.8: Robotic demonstration of the sequence generation architecture on a Khepera robot equipped with a color camera. (a-d) The robot navigates the arena, locating and approaching colored objects in the learned order.

This simple random search scenario demonstrates core properties of the DFT sequence generation architecture: (1) the capacity to derive and maintain stable representations of action goals from a simple sensory system, (2) the capacity to control real-world motor behavior, and (3) the capacity to obtain a reliable sensory signal that controls the switch from an action to its successor. Because it takes variable amounts of time to find an object of the requested color, this task highlights (4) the capacity of the sequence generation system to stabilize an action goal until it has been achieved. Finally, the learning phase demonstrates (5) the capacity to autonomously acquire the serial order of a sequence from sensory signals.

5.3 Robotic implementation

5.3.1 The action and condition of satisfaction systems

The action and condition of satisfaction fields work on the basis of fairly low level sensory information. Because color is the perceptual dimension that specifies which action within a sequence is currently activated, the action field is defined over hue as the dimension, x. A supra-threshold peak of activation in the action field provides ridge-shaped input to a perceptual color-space field defined below. Such input is localized along the color dimension but constant along space.

The condition of satisfaction field should generate a peak when a block is approached whose color matches the hue values currently activated in the action field. The condition of satisfaction field is, therefore, defined over the same hue dimension, labelled y when referring to the condition of satisfaction field. Formally, the mapping, $T(x, y)$, in Eq. (5.5) is the identity in this simple example. A peak in the action field centered over the hue value, x_0, provides localized input to the condition of satisfaction field around that same value. Perceptual input to the condition of satisfaction field is received from a central portion of the camera image. The dynamics of this field is tuned such that the detection instability is reached only when a sufficiently large blob of pixels register hue values that match the input from the action field.

5.3.2 The color-space field: Interface to sensors and motors

The two-dimensional color-space field (Figure 5.9A) plays a dual role. On input, this field performs a parallel visual search, generating a peak at a spatial location at which the hue of a contiguous blob of pixels in the camera image matches the localized hue input from the action field. On output, this field provides the target direction for the robot's movement toward the location at which the color match has been detected. The color-space field is defined over dimensions of hue and horizontal axis of the image plane (the horizontal spatial dimension is sufficient to control the vehicle's heading direction). Figure 5.9 shows how the sensory input to the color-space field is computed from the camera image: The distribution of hue values within a vertical column in the image, Figure 5.9B, is computed at each location, x_{im}, along the horizontal axis of the image plane, Figure 5.9C.

During sequence production, the color-space field is tuned such that a peak can be generated only when input from the camera overlaps sufficiently with ridge input

5. ORDINAL DYNAMICS BASED ON DISCRETE ACTIVATION VARIABLES

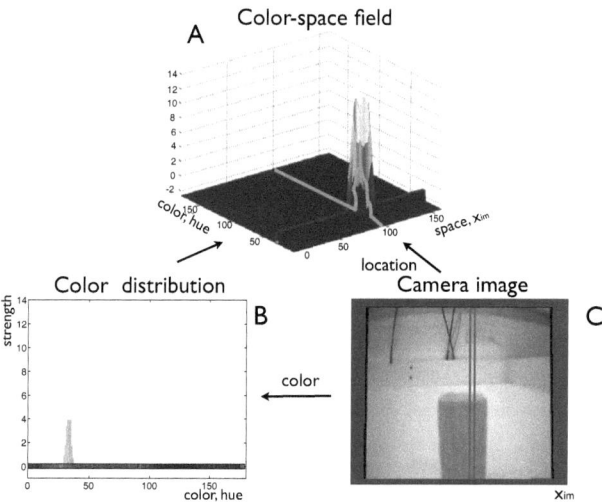

Figure 5.9: The perception of the robot: for each column in the camera image (C), the color distribution (B) is determined. This distribution forms input to the location in the perceptual color-space field (A) that corresponds to the position of the column in the image. A large enough blob contiguous in both color and space induces a peak of activation in the color-space field, which is the stabilized representation of the sought object in the image-based reference frame.

from the action field that represents the color currently searched for (see the ridge along the spatial dimension in Figure 5.9A). Sufficiently strong and spatially focused input from the camera, which falls onto the ridge, drives the field through the detection instability. The emerging peak signals that a suitably colored object has been found within the visual array. This peak then sets an attractor for a dynamics of heading direction that controls the robot's movement (see Appendix for details). As a result, the robot turns to and moves toward the location at which the suitably colored object has been detected. As the robot approaches the object, its visual projection onto the image plane grows in size until it triggers a peak in the condition of satisfaction field and the transition to the next color task in the sequence unfolds.

The strength of attraction toward the target direction scales with the size of the peak in the color-space field. The dynamics of heading direction also receives contributions

5.3 Robotic implementation

from distance sensors for obstacle avoidance. In the absence of a peak in the color space field, these obstacle avoidance contributions are thus alone in determining the robot's movement. In this mode, the robot will effectively wander around within an enclosed arena, generating what amounts to random search behavior. The velocity of the robot is controlled by a separate dynamics, which depends on the measured distance values, slowing the robot down in the vicinity of obstacles.

The perceptual color-space field is described mathematically by the dynamics of a two-dimensional neural field, Eq. (5.6), spanning the dimensions of color and horizontal position in the image. This field receives visual input from the camera of the agent: for each column in the image the color distribution is extracted and input to the corresponding row of the color-space field (see Figure 5.9 and Section 5.3.2 "The color-space field: Interface to sensors and motors").

$$\tau \dot{P}(x, x_{sp}, t) = - P(x, x_{sp}, t) + h_P + \int f(P(x, x_{sp}, t)) w(x - x', x_{sp} - x'_{sp}) dx' dx'_{sp} \\ + c_A^P f(U_A(x,t)) + c_{vis}^P V(x, x_{sp}, t),$$
(5.6)

where c_A^P is the constant characterizing strength of the input from the action field, $U_A(x,t)$, Eq. (5.3), that represents the ongoing action; $V(x, x_{sp}, t)$ is the visual input from the camera of the robot with amplitude c_{vis}^P. Other notation is the same as in Section 5.2 "DFT as a framework for embodied sequence generation". The interaction kernel $w(x - x', x_{sp} - x'_{sp})$ is a two-dimensional symmetrical Gaussian.

A large contiguous single-color blob in the image induces a localized peak of activation in the color-space field, which represents the colored object at the corresponding position in the image. The location of the peak along the spatial dimension specifies the contribution of the target to the dynamics of heading direction defined next.

During sequence learning, a homogeneous input (boost) to the color-space field ensures that the input from the camera alone is sufficient to induce a peak in the color-space field. Such a peak represents the detection and, in the presence of multiple colored objects, selection of a colored object. The color information is passed on to the action field. Specifically, for each hue value, x, the maximal value of activation in the color-space field along the spatial dimension, x_{sp}, is fed into the action field. In the learning mode, the central portion of the camera image also provides input to the condition of

5. ORDINAL DYNAMICS BASED ON DISCRETE ACTIVATION VARIABLES

satisfaction field. When a colored block is shown to the robot sufficiently close and centered on the robot's camera, the condition of satisfaction field generates a peak that inhibits the ordinal set, stopping strengthening of the synaptic weights. Notably, the sensory event that induces the condition of satisfaction signal must have a duration sufficient for a peak of activation to be built in this field. This duration is on the time scale of the dynamics (in milliseconds range in the current robotic implementation of the model). The peak in the CoS field during learning is released only when the perceptual input to this field ceases (when the teacher puts the colored block out of the camera view) because of the higher impact of the perceptual input during the learning phase. The transient to the next item in the sequence is thus accomplished and the next color's representation is associated with the next ordinal node.

5.3.3 Motor system

In the color-search task, the movement of the robot is controlled by a dynamics of the heading direction of the robot, Eq. (5.7) (Schöner et al., 1995). The desired change of heading direction, $\Delta\theta = \dot{\theta}(t)\Delta t$, at each time step, t, is translated into a difference of commands for the two wheels of the robot and causes the desired rotation. The new heading direction, θ, is never computed explicitly here, but the robotic hardware, as part of the loop, "integrates" the dynamics.

$$\tau_{nav}\dot{\theta}(t) = \lambda_{obs}F_{obs,i}(t) + \lambda_{tar}F_{tar}(t). \qquad (5.7)$$

Here, λ_{tar} and λ_{obs} are constants controlling the strength of the target (attractor) and obstacles (repellors) contributions. The functions, $F_{obs,i}$ and F_{tar}, describe the contributions to the dynamics from obstacles and the target perceived by the robot. The obstacle contribution is calculated with Eq. (5.8).

$$F_{obs} = \sum_i S_i(t) \cdot \exp\left[-\frac{\Psi_i}{2R_{obs,i}(t)^2}\right]. \qquad (5.8)$$

Here, $S_i(t)$, $i = 1..8$, are the signal strengths and Ψ_i are angular locations relative to the heading direction of the eight infrared sensors of the robot. $S_i(t)$ are calculated according to Eq. (5.9). $R_{obs,i}(t) = \arctan[\tan(\frac{\Delta\Psi}{2}) + \frac{R}{R+D_i(t)}]$ is the angular range of an obstacle, that depends on the size of the robot, R, opening angle of the sensor, $\Delta\Psi$, and distance to the obstacle, $D_i(t)$. $D_i(t)$ is the distance to an obstacle in direction of the

5.3 Robotic implementation

i^{th} sensor that is calculated from the sensor's value through an empirical calibration function, $D_i(t) = 0.9 \cdot \exp -0.007 \cdot Ir_i(t))[mm]$, where β_1, β_2 are numerical constants.

$$S_i(t) = \beta_1 \cdot \exp\left[-\frac{D_i(t)}{\beta_2}\right]. \qquad (5.9)$$

The target contribution is calculated by extracting the spatial information from the output of the color-space field, $P(x, x_{sp}, t)$, Eq. (5.6). Thus, the target contribution is proportional to a distance from the peak of activity in this neural field to the center of the field, both calculated for the spatial projection, $P_{pr}(x_{sp}, t)$, of the two-dimensional neural field's output, Eq. (5.10).

$$F_{tar}(t) = \int_0^N P_{pr}(x_{sp}, t) \left(x_{sp} - \frac{N}{2}\right) dx_{sp}. \qquad (5.10)$$

Here, N is the (arbitrary) size of the perceptual color-space field. The projection $P_{pr}(x_{sp}, t) = f\Big(P(argmax_x(P(x, x_{sp}, t)), x_{sp}, t\Big)$ is a projection of the output of the color-space field on the spatial dimension, x_{sp}; x is the color dimension of the color-space field. Thus, integrating the Eq. (5.7), the robot minimizes the angular distance between the current heading direction, θ, and the direction to the target, θ_{tar}, both of which, however, are not calculated explicitly.

The dynamics, Eq. (5.7), integrates impacts of the two sensor modalities – color vision and infrared sensing. The color vision is responsible for target acquisition and is in turn affected by the dynamics of sequence generation. The sensori-motor dynamics, Eq. (5.7), thus provides for a low-level behavior organization.

The speed of the robot is controlled by Eq. (5.11):

$$\tau_{vel}\dot{v}(t) = -v(t) + \left(1 - \lambda_{obs}^v(t)\right) \cdot v_{tar} + \lambda_{obs}^v(t) \cdot v_{obs}, \qquad (5.11)$$

where $v_{obs} > v_{tar}$ are constant velocities with and without an obstacle in the vicinity of the robot respectively; λ_{obs}^v is a switch, that turns non-zero if an obstacle is perceived by an infrared sensor; τ_{vel} is the time-constant of the dynamics.

5. ORDINAL DYNAMICS BASED ON DISCRETE ACTIVATION VARIABLES

5.4 Results: the model in action.

5.4.1 Sequence learning

Figure 5.10 illustrates how the robot is taught a sequence of colors. When the first block is presented to the robot's camera (Figure 5.10, first column), the prevalent color in the image induces an activation peak in the perceptual color-space field. This peak is projected onto the color dimension of the action field and induces a peak there. Note that this reversal of the direction of coupling compared to the sequence production mode is brought about merely by the boost of activation to the perceptual color-space field. The presence of a peak in the action field, while a particular ordinal node is active, strengthens the synaptic connections from that node to the activated sites of the action field.

The activation peak in the action field also provides localized input to the condition of satisfaction field. Therefore, as soon as perceptual input to the condition of satisfaction field is sufficiently strong – because the object is brought sufficiently close to the camera – an activation peak is induced in the condition of satisfaction field. This peak then inhibits the ordinal set (Figure 5.10, second column), leading to activation in the ordinal system to drop below threshold. The strengthening of synaptic weights stops at this point. When the teacher removes the colored block from the robot's view (Figure 5.10, third column), the activation peaks in the action and the condition of satisfaction fields decay completely, releasing the ordinal set from inhibition. Due to the dynamics of the ordinal and memory nodes, the next ordinal node becomes activated.

When the next block with the next color is presented to the robot, the new color is detected in the color-space field (Figure 5.10, last column) and passed on to the action field in the same fashion. Synaptic connections between the next ordinal node and the activated sites of the action field are strengthened. The procedure continues until all color blocks have been presented to the robot. The outcome of the learning procedure is a set of learned connection weights from the ordinal nodes to the action field, which may now direct the flow of activation during sequence generation.

5.4.2 Sequence production

Figure 5.11 illustrates how sequential color search behavior is produced by the robot. An active ordinal node induces an activation peak in the action field through the learned

5.4 Results: the model in action.

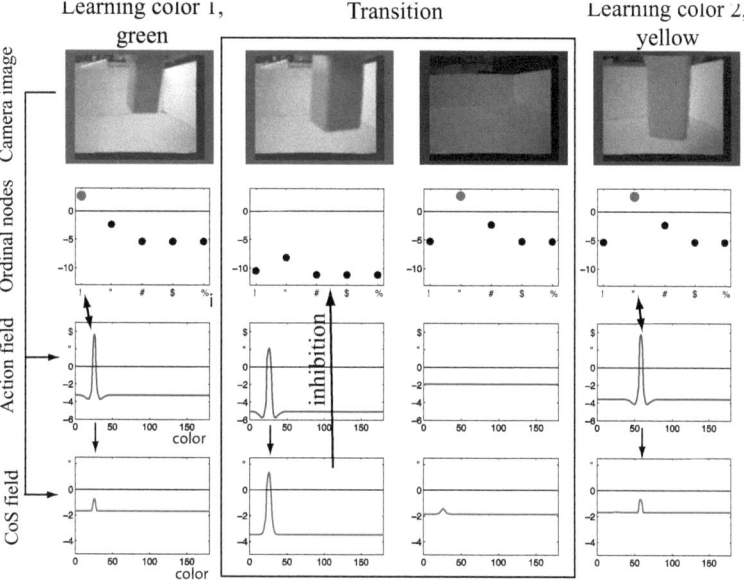

Figure 5.10: Sequence learning in the DFT architecture driven by the visual input presented by a user. The sequential transition between two subsequent colors is explained in the main text.

5. ORDINAL DYNAMICS BASED ON DISCRETE ACTIVATION VARIABLES

synaptic connections (Figure 5.11A). The location of this peak specifies the color that must now be searched (here, "green"). The localized output of the action field sends ridge-shaped input to the perceptual color-space field (Figure 5.11B), facilitating activation of that field at "green" sites. Thus, the locations of green objects in the camera image (Figure 5.11C) compete for activation in the color-space field. A location receiving input from the largest green object surpasses the activation threshold first. Lateral inhibitory interaction within the color-space field selects and stabilizes the representation of this object (Figure 5.11B). This activation peak in the color-space field now controls the robot's movement by specifying an attractor in its heading direction dynamics. As the robot moves, the shifting location on the sensory surface of the visual location of the selected green object is tracked by the peak in the color-space field.

The activation peak in the action field pre-activates the condition of satisfaction field (Figure 5.11D), making it sensitive to "green". When the robot has approached the target block, the associated color blob takes up a large portion of the image (Figure 5.11H). As a result, perceptual input to the condition of satisfaction field, summed with the prior input from the action field, surpasses the activation threshold of the detection instability (Figure 5.11I). The emerging activation peak in the condition of satisfaction field signals the successful accomplishment of the color-search action at this stage of the sequence.

The peak in the condition of satisfaction field inhibits the ordinal nodes, triggering the cascade of transition instabilities. The peak representing the current action in the action field decays, which causes decay of the peak in the condition of satisfaction field. Ultimately, the representation of the next action in both ordinal and action systems is activated. The previous color loses its advantage in the color-space field, where the peak decays. Consequently, the attractive force in the heading direction dynamics wanes. The green block now only acts as an obstacle, helping the robot to get on its way toward searching for the next color.

The Figure 5.12 shows the time-course of the dynamics of the ordinal nodes, the action field, the condition of satisfaction field, and the spatial projection of the color-space field during learning and production of a sequence "red-blue-green-blue-yellow" (RBGBY).

In the following, we present several runs of the robot system in different environments, merely varying which sequence of colors is taught during sequence learning and

5.4 Results: the model in action.

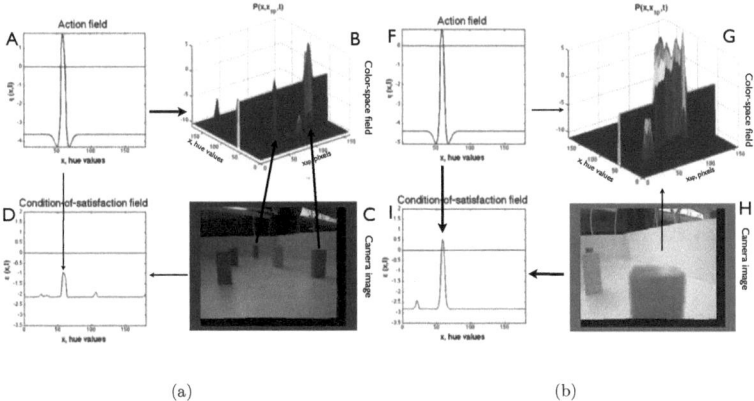

Figure 5.11: Two snapshots of dynamics of the DFT sequence generating architecture during sequence production on a robot. (a) Action "Find green" is active, the robot approaches one of the two green objects in view. (b) Condition of satisfaction is activated by the input induced by a close green object centered in the camera image.

5. ORDINAL DYNAMICS BASED ON DISCRETE ACTIVATION VARIABLES

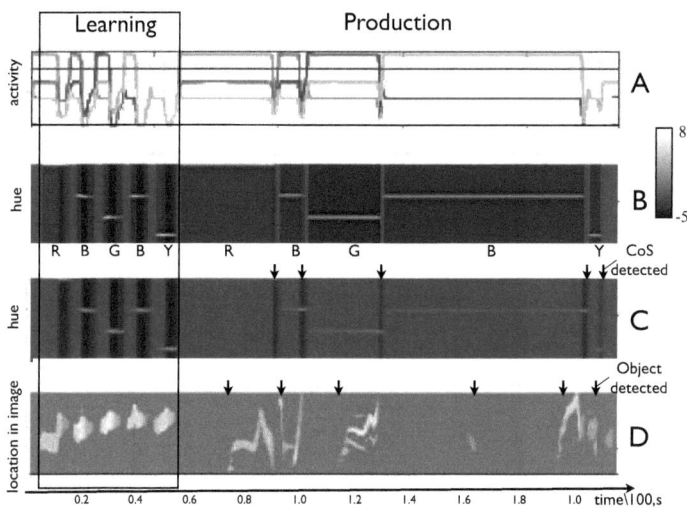

Figure 5.12: Time course of a robotic demonstration during learning and production of a sequence "find red-blue-green-blue-yellow". **A**: Activations of five ordinal nodes. **B**: Activation of the action field. Activation is above threshold in localized peaks and negative otherwise. **C**: Activation in the condition of satisfaction (CoS) field. Arrows above the plot mark time-points, at which the CoS field was activated (detection instabilities). **D**: A projection of the perceptual color-space field onto the spatial dimension (horizontal axis of the image plane). Light regions correspond to positive, supra-threshold, activation, dark regions denote negative activation. Arrows mark time-points, at which the object of interest in each ordinal position first appeared in the visual array of the robot. The wandering behavior changed to "approach target" behavior at these points.

5.4 Results: the model in action.

the physical arrangement of objects in the arena during sequence production. Exactly the same parameter setting is used for the neural dynamics in all demonstrations.

5.4.3 Timing of actions

The core property of the DFT architecture is the stability of action representations at each ordinal position enabling the system to tolerate variable durations of the individual actions when the sequence is executed in an unknown environment. In the simulations illustrated in Figure 5.13, the durations of actions range from a few seconds to two minutes due to different spatial arrangements of obstacles and targets.

5.4.4 Flexibility of sequence generation: no problems with repetitions

The DFT architecture does not make use of direct connections between the representations of actions at different stages of the sequence (no "chaining" mechanism), nor is inhibition of the previous items essential for sequence learning and production. This makes it possible to learn any sequence, including those in which the same actions are repeated at different, even adjacent ordinal positions. This is demonstrated by the run illustrated in Figure 5.13(a), in which the "blue" is searched at the second and fourth step. In the run illustrated in Figure 5.13(b) the same color "green" is requested twice in a row. Disentangling ordinal information from the content of each action enables this form of flexibility.

5.4.5 Noisy environments

To illustrate the robustness of sequence generation in the DFT architecture, some sequences were acquired outside the arena. Thus, during the learning phase input was more complex, no longer dominated by a single prevalent color. Figure 5.13(c) shows that the system was still able to detect the most salient color during the learning phase and to learn the correct sequence.

The robotic demonstrations reported here are illustrative. Each demonstration was repeated several (5-10) times with the same configuration of colored blocks, and also with different configurations, and with different sequences to ensure robustness of the architecture. Still, because of the simplicity of the sensory-motor system, chosen on

5. ORDINAL DYNAMICS BASED ON DISCRETE ACTIVATION VARIABLES

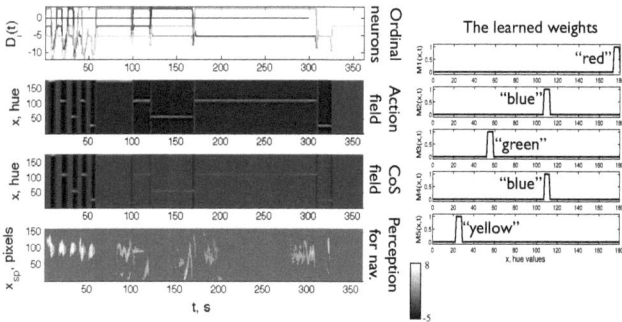

(a) Sequence "red-blue-green-blue-yellow" acquired and produced in a robotic arena.

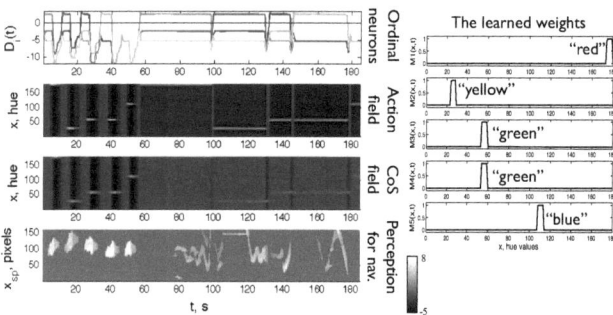

(b) Sequence "red-green-green-blue-yellow" acquired and produced in the arena.

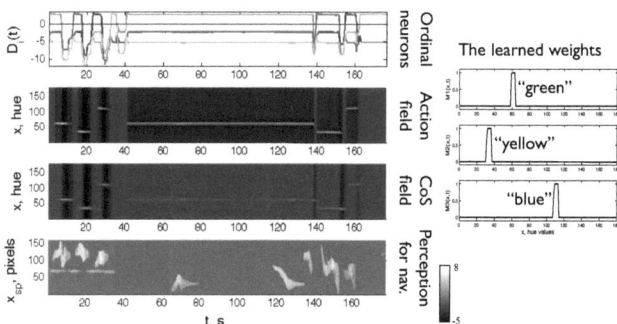

(c) Sequence "green-yellow-blue" acquired outside and produced in the arena.

Figure 5.13: Three demonstrations of sequence learning and production. **Left:** The time-courses of activation in the functional modules of the DFT sequence generation architecture. **Right:** The synaptic weights holding the sequence of colors after learning.

purpose here to demonstrate that the cognitive dynamics can be coupled to the low-level sensory-motor dynamics, the system is not able to deal with an unconstrained environment. In such environment, the problems of temporal segmentation of events and spatial segmentation of objects have to be addressed on the level of perceptual modules. A more complete architecture to represent objects and scenes (Faubel & Schöner, 2008) must replace the simple color-space perceptual field in order to enable robust functioning of the system in complex feature-rich environments.

5.5 Discussion

Stability of the functional neural states is the critical property of the model that enables the agent to connect to noisy and unreliable sensory information. Stability is in conflict with the need to transition among the sequential stages and is, therefore, absent from most neural dynamic models of sequence generation (Botvinick & Plaut, 2006; Bradski et al., 1994; Deco & Rolls, 2005; Rabinovich et al., 2006). In the DFT model presented here, I overcome this conflict by introducing the concept of a condition of satisfaction, a stable neural representation of the terminal state of an action. This mechanism makes it possible to execute sequences in environments in which the durations of individual actions are unpredictable and the signal for a sequential transition must be extracted autonomously from noisy sensors.

Neural inspiration for the extended model comes from findings of a separate neural substrate for the ordinal position of an element in a sequence, along with the neural representation of the motor characteristics of the associated movements (Aldridge & Berridge, 1998; Procyk & Joseph, 2001). The condition of satisfaction system may be thought of as a potential reward mechanism. Together with the ordinal nodes, the condition of satisfaction system resembles the mechanisms of action selection hypothesized to reside in the basal ganglia (Redgrave et al., 1999), which project onto the distributed neural representations of actions and perceptual states. In the DFT sequence generating model, the adjustable neural weights hold the memory for each element of a sequence. These weights are established from a single exposure to a sequence element. Such fast learning is possible due to the explicit segregation of ordinal position within the ordinal layer and stands in contrast to the extensive learning required in models of

5. ORDINAL DYNAMICS BASED ON DISCRETE ACTIVATION VARIABLES

serial order that are based on distributed and overlapping representations of the order along with motor and perceptual features (Elman, 1990).

The modified architecture is more neurally plausible, the computational resources are used more sparsely here than in the previous model. The dynamics of ordinal nodes can be analyzed in separation from the dynamics of neural fields, which allows to better understand the dynamics of sequential switching, in particular, the cascade of instabilities that leads to a transition between two stages of a sequence. The extension of the model enables the condition of satisfaction system to represent graded aspects of actions that are specified by graded sensory information controlling the sequential transitions during learning and production. The robustness of the system to variations in action duration, to perturbations in visual input, as well as the autonomy and embodiment of the model are demonstrated in robotic experiments.

In the following Chapter, I proceed to exploit the solid basis established in the modified DFT sequence generation architecture to learn and produce heterogeneous sequences.

6

A multi-dimensional DFT sequence generation architecture

6.1 Motivation for the multidimensional DFT sequence generation architecture

In the color search example used in the previous chapters to demonstrate the core properties of the DFT sequence generation architecture, only one characteristic variable, spanning a single dimension of a dynamic neural field, described all possible actions. In order to demonstrate that this is not a principled limitation of the model, I introduce in this chapter an implementation of the model for a robotic scenario with heterogeneous actions. Thus, the Khepera robot is equipped with a gripper now, which can be raised or lowered and closed or opened. The task for the robot is now to search for particular colors, stop in front of a block of the sought color, grasp it and transport to a different color block. As I want to avoid considerations of the behavior organization here, the sequence of the needed combination of color search actions, lowering of the arm and closing or opening of the gripper is demonstrated by the user and memorized in the sequence generating architecture. The robot performs imitation rather than planning here. In the learning process, the teacher triggers the needed actions on the robot, or puts the color blocks in front of the camera, the perceptual systems of the robot and the sequence generating dynamics make sure that the correct sequence of actions in different modalities is stored in the connection weights between ordinal nodes and dynamic neural fields. During production, the robot performs the learned sequence

6. A MULTI-DIMENSIONAL DFT SEQUENCE GENERATION ARCHITECTURE

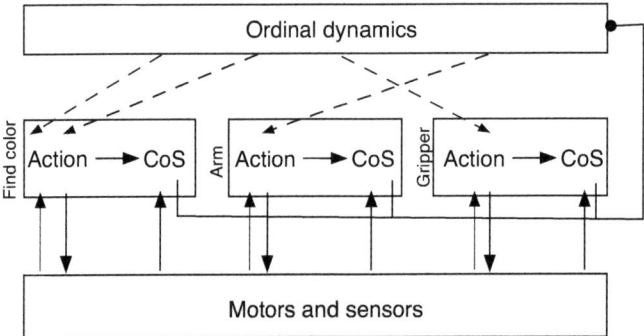

Figure 6.1: The sequence generating architecture including action systems specific to the robotic implementation.

autonomously in an arena, relying on its own sensors.

6.2 The architecture

In order to represent heterogeneous sequences in the DFT sequence generation architecture, the ordinal system is connected through modifiable synaptic weights with several action fields, each coding a particular action modality. The connection weights constitute the memory for the sequence and are learned within a single demonstration of the action. Localized peaks in the action fields impact on the motor system of the agent, shaping attractors of the motor dynamics, and thus specifying a particular action. The stability of the neural field dynamics guarantees that the impact on the motor dynamics is sustained as long as needed for the current action goal to be achieved. For each action modality, a condition of satisfaction (CoS) neural field is defined. Localized input from the active regions of action fields makes the corresponding CoS field sensitive to sensory input signaling successful accomplishment of the action. Positive activation in the CoS field triggers a cascade of instabilities that bring about the sequential transition to the next action.

An overview of the architecture for the present robotic implementation is given in Figure 6.1. In this Section, I briefly describe the dynamics of the three main parts of the architecture, relegating the detailed mathematics to Section 6.3.

6.2 The architecture

6.2.1 Ordinal nodes

The ordinal system of a heterogeneous DFT sequence generation architecture has the same dynamics and structure as described in Chapter 5.

The active node now provides input to one or several of the action fields through modifiable connection weights, depending of whether one or several actions where demonstrated simultaneously during learning. When the final state of the action is reached, the ordinal set is inhibited by input from the condition of satisfaction system. If several action fields are active, activation of both their condition of satisfaction fields is needed to deactivate the ordinal system.

6.2.2 Action fields and synaptic projections

The action fields are dynamic neural fields defined over characteristic dimensions of actions. A self-stabilized peak in every action field is induced from the ordinal nodes and controls the robotic action by setting attractors for the sensory-motor dynamics.

Action fields also receive perceptual input about the characteristic action parameter of an ongoing or demonstrated action. This input is critical during sequence learning, when it induces a localized, self-stabilized peak of activation in the action field that represents a demonstrated action. The connection weights linking the active ordinal node to the region of positive activation of the action field are upweighted according to a Hebbian learning rule.

6.2.3 Condition of satisfaction fields

For each action in a sequence, its condition of satisfaction (CoS) is defined. The CoS dynamic fields are spanned over metric dimensions that characterize the terminal states of the actions. The CoS fields receive localized inputs from active regions of the action fields. The mapping between the action fields and CoS fields could be learned from previous experience, but was assumed given in the current implementation. The action field input preactivates its CoS field so that the CoS is sensitive to perceptual input that matches the expected terminal state of the action. When such a terminal state is detected, a self-stabilized peak in the CoS field is induced, that then inhibits the ordinal system. This suppresses activity in the ordinal system and thus removes the input the ordinal system provided to the action fields. As a result, the peak in the

6. A MULTI-DIMENSIONAL DFT SEQUENCE GENERATION ARCHITECTURE

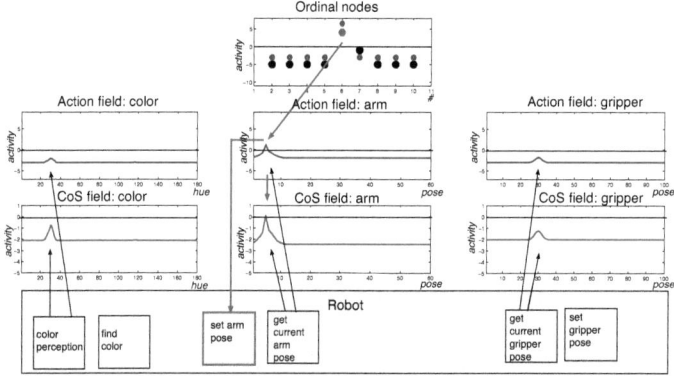

Figure 6.2: Snapshot of the dynamics of the multimodal sequence generating model: an action "lower arm" is active, the "color" and "gripper" action and condition of satisfaction (CoS) fields receive only subthreshold perceptual input. See text for details.

action field becomes unstable and decays. This removes the previous input to the CoS field, pushing that system through the same instability and leading to the decay of the peak in the CoS field. This transition, finally, removes inhibition from the ordinal set, and the next ordinal node becomes activated. This cascade of instabilities separates the different ordinal positions in a sequence both during sequence learning and production.

6.3 Mathematical description of the model

The dynamics of the ordinal nodes and the corresponding memory nodes are described by the equations Eq. (5.1) and Eq. (5.2) from Chapter 5.

The dynamics of the stack of *action fields* follow the equation:

$$\tau \dot{U}_j^A(x_j, t) = - U_j^A(x_j, t) + h^A + \int f\Big(U_j^A(x'_j, t)\Big) w(x_j - x'_j) dx'_j \\ + \sum_{i=1}^{N_d} f\Big(d_i(t)\Big) M_i(x_j, t) + c_p^A I_p^A(x_j, t) \tag{6.1}$$

Here, the first three terms define the generic neural field dynamics with a negative resting level h^A, and the lateral interactions term. N_d is the number of ordinal nodes, activities of which, $d_i(t)$, are thresholded by a soft sigmoid function $f(\cdot)$ and then projected on the action field through the modifiable weights, $M_i(x_j,t)$, where j numbers the actions fields, and i numbers the ordinal nodes. c_p^A controls the strength of the perceptual input, $I_p^A(x_j,t)$, essential during sequence learning.

When an ordinal node is active (i.e. $f(d_i(t)) > 0$), its activation propagates to the action field, providing a localized input to this field. The shape of this input is defined by the *synaptic weights*, $M_i(x_j,t)$, which are modified during sequence learning according to a Hebbian-like rule:

$$\tau_l \dot{M}_i(x_j,t) = \Big(- M_i(x_j,t) + f\big(U_j^A(x_j,t)\big)\Big) \cdot f\big(d_i(t)\big) \qquad (6.2)$$

The *CoS neural fields* evolve according to a dynamical equation:

$$\tau \dot{U}_j^C(y_j,t) = - U_j^C(y_j,t) + h^C + \int f\Big(U_j^C(y_j',t)\Big) w(y_j - y_j') dy_j' \\ + \int T(x_j, y_j) f\Big(U_j^A(x_j,t)\Big) dx_j + c_p I_p(y_j, t). \qquad (6.3)$$

Here, the activity of the jth CoS field $U_j^C(y_j,t)$ is defined over the neural dimension, y_j. The transfer function $T(x_j, y_j)$ defines the mapping between the dimensions characterizing actions and their terminal states ($T(x_j, y_j) = \mathbf{1}$ in the implementation presented here). Positive activation in the action field (where $f(U_j^A(x_j,t)) > 0$) propagates to the CoS field through this mapping. The constant c_p controls the strength of the perceptual input $I_p(y_j, p)$, h_C is the resting level, and τ is the time-constant of the field's dynamics.

6.4 Robotic implementation

In order to demonstrate learning and production of action sequences in an embodied setting, I describe an exemplary implementation of the architecture on a mobile robot vehicle of the Khepera type.

6.4.1 The robotic scenario

In the learning phase of the scenario, a teacher demonstrates arbitrary sequences of alternating movements or colors to the robot. After learning, the robot is put in an arena in which colored blocks have been distributed. The first node of the ordinal set is activated by a "go" signal and the robot performs the learned sequence.

In this scenario, three action modalities are considered: search of objects of a given color, lifting movement of the robotic arm, and opening of the robotic gripper. The

6. A MULTI-DIMENSIONAL DFT SEQUENCE GENERATION ARCHITECTURE

Figure 6.3: The robotic scenario: a sequence "find yellow"-"lower gripper"-"close gripper"-"find green"-"lower gripper"-"open gripper" results in the robot transporting a yellow object and depositing it on a green object.

three metric parameters, characterizing the three action modalities - color, arm height and gripper opening - span the three neural dimensions over which the action fields are defined.

The particular sequence, described in the following is "find green - lower arm - close gripper - lift arm - find yellow - lower arm - open gripper", which results in the grasping of a green block, its transportation to a yellow block and its depositing there (Figure 6.3).

6.4.2 Implementation

6.4.2.1 Teacher interaction

During learning, graded sensory input specifies the demonstrated action. Additionally, the teacher specifies the intended action modality by pressing buttons in a GUI. That signal provides a homogeneous activation boost that raises the resting level of the cue action field. Thus, for instance, specifying "color" boosts the color action field so that the color of the block that is currently present in the robot's visual array induces a localized peak of activation. Similarly, specifying "arm" boosts the arm-pose field and a peak is built at the location that represents the sensed current elevation of the robot's arm. Specifying "gripper" activates the gripper field in the same manner.

6.4.2.2 Perception and motors

The robot's color camera provides perceptual input to the color-search modality. Color extraction is performed through a color-space neural field, which models early stages of visual processing (see Chpater 5 that employs the same method). The search for colored objects is accomplished by an attractor dynamics of the heading direction of the robot, as described in (Schöner et al., 1995).

The arm of the Khepera robot can be lifted and lowered. This movement is controlled by a one-dimensional neural field defined over the dimension "arm elevation". For a robotic arm with several degrees of freedom, the corresponding field can be defined over the three spatial dimensions that specify the spatial position of the end-effector. A localized peak of activation in the arm field sets an attractor for the dynamics of arm movement. During learning, the sensed current position of the arm sets a localized sensory input to the arm action field. That input may induce a peak if the teacher specifies "arm" as the action modality to attend to.

The robotic gripper consists of two bars which can be closed or opened. The action field is defined over the dimension "gripper opening". A peak in that field initiates either the "closeGripper()" or the "openGripper()" commands (only these two options exist for the Khepera's gripper), depending on where along that dimension the peak is centered relative to the current gripper position. During learning, a graded sensory signal reflecting the current opening of the gripper is fed into the action field inducing a peak, whose location represents the action at that ordinal position.

In this implementation, the CoS fields for the color, arm and gripper modalities are defined over the same dimensions of color, arm elevation and gripper opening. For color, the sensory input to the field is derived from the central portion of the robot's field of view; consequently, the CoS field is activated when an object of the searched color is close to the robotic camera and centered in view. The arm and gripper CoS field are activated when a match is detected between the attractor position for the corresponding movements and the actual position of the arm and the gripper.

Figure 6.2 shows a snapshot of the sequence generation dynamics. The fifth ordinal node is active here and projects onto the "arm" action field through the neural connections. The ordinal input induces an activity peak in the action field, and the arm elevation value $pose \approx 5$ is set as an attractor for the arm movement. When the CoS field detects the arm elevation of $pose \approx 5$, an activity peak arises in the CoS field and starts to inhibit the ordinal nodes. The "color" and "gripper" action and CoS fields also receive perceptual input from sensors of the robot here, but this input alone is not sufficient to induce peaks in those fields.

6. A MULTI-DIMENSIONAL DFT SEQUENCE GENERATION ARCHITECTURE

6.4.3 Results

Figure 6.4 presents the time courses of activity of the eight ordinal nodes and the three action fields during learning (Figure 6.4(a)) and production (Figure 6.4(b)) of the sequence "find green - lower arm - close gripper - lift arm - find yellow - lower arm - open gripper". During learning, the teacher demonstrates the actions to the robot by naming the modalities of interest and triggering the needed actions. The demonstrated actions are detected in the action fields at times that are marked with arrows on the Figure 6.4(a). Lightly shaded regions mark regions in the action fields with positive activity. These regions are associated with the currently active ordinal node by quickly adaptable neural weights. The CoS is activated shortly thereafter by perceptual input that signals the accomplished action. The CoS signal inhibits the ordinal set (top row on the Figure 6.4) and triggers an instability, after which the next ordinal node is activated. The next action, demonstrated by the teacher is associated with this node.

After learning, the robot is put into the arena, in which colored objects may be distributed in a different spatial arrangement. A "go" signal brings the robot into its initial state (gripper is high and opened) and activates the first ordinal node. A stable peak of activation emerges in the "color" action field (dark red region on the Figure 6.4(b)). This peak represents the color-search action for however long it takes to locate, center and approach an object of the specified color (with $hue \approx 100$, green, here). The CoS field detects that an object of the specified color is sufficiently close and centered, builds a peak, and inhibits the ordinal set. This triggers the ordinal transition, a cascade of instabilities leading to the activation of the second ordinal node. The second action, "arm lowering" is performed by the robot, its termination again being controlled by the CoS field and pertinent sensory input. The rest of the sequence is acted out in the same manner. The robot picks up the green block and delivers it to the yellow one.

Note the flexible timing of the actions both during learning, when the timing is controlled by the teacher, and during production, when the duration of actions depends on the current situation in the arena.

In experiments, in approximately 10% of trials the robot fails to accomplish the sequence, because the small block slides out of the gripper, or because of a wrong positioning in front of the object, on which the small block is to be deposited. In this work, the trial was interrupted if such a failure happened. Autonomous detection of failures and initiation of a correcting sequence can be accomplished in behavior organization architecture and are outside the scope of this work.

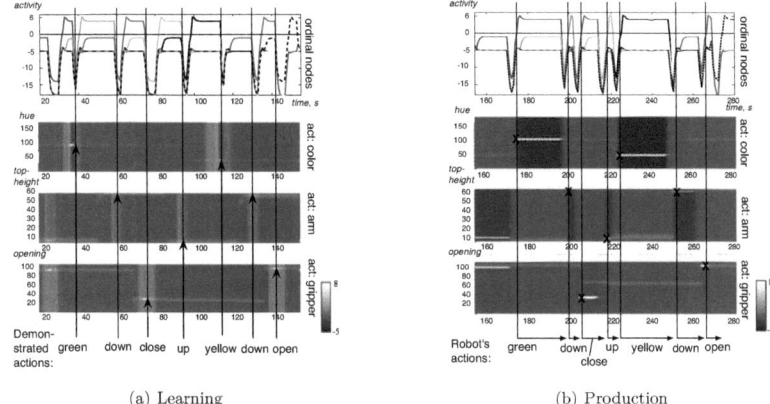

Figure 6.4: Time-courses of the dynamics of ordinal nodes and the three action fields during learning and production of a multimodal sequence. Dark regions on the field plots correspond to a negative activation in neural fields, light stripes with a dark midline are traces of peaks of positive activation, light regions with a lighter gray around them are traces of localized subthreshold preshapes.

6.5 Discussion

In this Chapter, I tested the extended DFT model for generating sequences of heterogeneous actions, that involve different motor and perceptual modalities. This generalization is an important step toward an approach to sequential behavior in natural settings in agents with a rich behavioral repertory.

With the introduction of multiple motor modalities, the sequence generation model touches on the problem of behavioral organization, the coordination of multiple different behaviors in time. At this point, the model sidesteps this issue. Within the model, different motor modalities do not interfere or interact. In the implementation, the three action systems link to different effector systems and degrees of freedom of the robot. Their simultaneous initiation, represented by co-existing peaks in the different action fields, is allowed. The actions "move gripper", "close gripper" and "search for red", for instance, can all be active at the same time. Integrating principles of behavioral organization would require a form of hierarchical organization, in which actions within a sequence may consist of an organized ensemble of sub-actions. The fundamental

6. A MULTI-DIMENSIONAL DFT SEQUENCE GENERATION ARCHITECTURE

mechanism established here for sequence generation in a stable neuronal dynamics may also provide a framework for how heterogeneous sequences of actions may be produced that lead to a given goal.

7

Application to control the cognitive dynamics within a DFT architecture for spatial language.

7.1 Spatial language: Motivation for the application.

Every day people use spatial terms like "in front", "right", and "behind" to describe the locations of real objects in the world. These spatial terms can be flexibly applied to objects with a variety of continuous, non-spatial features such as color or shape. This capacity to integrate spatial relations, spatial language, and continuous object features to structure behavior in the real world is central to human communication and coordinated action. Effective spatial language is also a necessary component of efficient human-robot interaction: spatial language provides a natural means by which humans communicate about and coordinate behaviors within a shared workspace. An embodied and grounded approach to spatial language provides not only the structure to use spatial language in communication, but also links spatial language to the sensory processes from which information about a scene is derived as well as to the behavioral processes through which object-oriented action may be engaged (Prinz & Barsalou, 2000).

The complexity and richness of spatial cognition and spatial language (Levinson, 2003) is an obvious challenge to such an approach. Comprehending or producing spatial language about a visual scene, for example, not only involves a neural scene representation that emerges from the retinal image but also the integration of long-term memory about objects and their features and the neural representation of spatial semantic terms

7. APPLICATION TO CONTROL THE COGNITIVE DYNAMICS WITHIN A DFT ARCHITECTURE FOR SPATIAL LANGUAGE.

(e.g. right, above, etc). Critically, these semantics must be applied to the current scene and they are often aligned with a reference object (Landau & Jackendoff, 1993).

Having these considerations in mind, a DFT architecture that accounts for human spatial language behaviors was introduced (Lipinski *et al.*, 2009e). Several colleagues and myself have extended this model and implemented it on the robotic platform CoRA (Lipinski *et al.*, 2009a,b,d; Sandamirskaya *et al.*, 2010). The extended architecture could successfully generate spatial language behaviors in interaction with a human user on a shared tabletop workspace, as well as account qualitatively for some empirical findings about human spatial language behaviors in more complex settings than the original model.

In order to perform a quantitative comparison of the model's behavior with experimental findings, the robotic model was modified to reflect more precisely the neurophysiology of visual processing (?). In the DFT spatial language architecture, a reference field holds spatial information about the location of the reference object. Its visual input is algorithmically preprocessed. In the modified architecture, this preprocessing is replaced by a direct connection to a perceptual color-space field. In analogy to the reference field, a target field is defined in the modified architecture. This activation field is defined over spatial dimensions and is directly connected to the color-space field. Because both target and reference fields now receive input from the perceptual color-space field, a *sequence of boosts* is needed to sequentially gate input from the color-space field to the spatial fields. In earlier work that was aimed at replicating experimental statistics on spatial language behaviors (?) this sequence of boosts was introduced manually. To enable autonomous sequencing of boosts, the sequence generation architecture can be used to control the sequential switching of dynamic regimes of the functional components of the spatial language architecture. Apart from introducing autonomy to the system, this enables comparison with experimental findings of the temporal structure of spatial language processes.

This chapter is organized as follows: in Section 7.2, I briefly present the robotic spatial language architecture. In Section 7.3, the modified spatial language architecture is introduced, in which a sequence of boosts is needed to serialize perceptual input to different functional parts of the architecture. In Section 7.5, I introduce the sequence generation mechanism to the modified architecture and also demonstrate the resulting autonomous behavior of the combined model. A brief discussion closes this chapter, and herewith the main part of the thesis.

7.2 Spatial language architecture: robotic implementation.

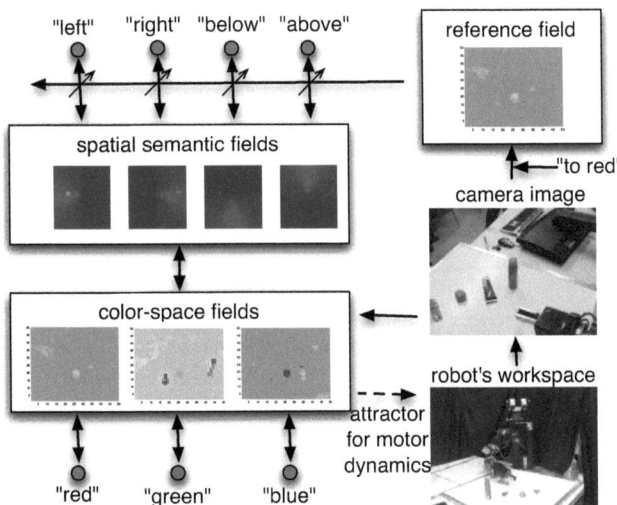

Figure 7.1: Overview of the DFT spatial language architecture. In the plots of *spatial semantic fields*, dark regions correspond to strongly negative activation of the neural fields, light gray regions – to less negative regions that receive spatial templates input mediated by the activity of spatial nodes and output of the reference field. In the plot of the *reference field*, lighter regions correspond to higher activation, and in the plots of *color-space fields*, dark regions correspond to high levels of activation in the middle plot (red originally), or to strongly negative levels of activation in the right plot (blue originally). The latter region stems from the negative input from the reference field. The architecture is explained in the main text.

7.2 Spatial language architecture: robotic implementation.

This Section outlines the overall structure and dynamics of the DFT spatial language architecture (Figure 7.1). Along with dynamic neural fields, discrete dynamical neural nodes are introduced to represent the linguistic input provided by the human user, as well as the linguistic output of the robot.

The robotic camera provides input to a set of dynamic neural fields that represent visual space through associations of colors and their locations. Each of these *color-*

7. APPLICATION TO CONTROL THE COGNITIVE DYNAMICS WITHIN A DFT ARCHITECTURE FOR SPATIAL LANGUAGE.

space fields (see Figure 7.1) receives input from pixels in the camera image, the hue value of which falls within a certain range, corresponding to a basic color (e.g. blue, red, green, etc). The resolution of color is low here because only a few colors are needed to represent the objects. Thus, several discrete two-dimensional spatial fields, each assigned to a hue interval, are sufficient to sample the visual scene. In principle, the continuum of colors could be resolved to an arbitrary degree of granularity in a three-dimensional implementation of the color-space field.

The visual input from the camera alone is not sufficient to activate the color-space fields. Language input specifying the color of the object boosts the resting level of the corresponding color-space field ("green" in Figure 7.1). When summed with the visual input from the camera, this activation boost induces an instability that leads to the formation of a single peak of activation centered over the object that provides the strongest input of the specified color. The spatial language input also influences the color-space fields' dynamics through the spatial semantic fields (see below). A peak of activation in a color-space field sets an attractor for the dynamics that controls the robotic movements.

The *reference field* (Figure 7.1) is a spatially tuned dynamic field, which also receives visual input. When the user specifies the reference object's color, the corresponding "reference-color" neural node becomes active and pixels with the specified color in the camera image provide input to the reference field. A reference field activation peak specifies the location of the reference object and continuously tracks its position. It also filters out irrelevant inputs and camera noise, thus stabilizing the reference object representation. Having a stable, but nonetheless updatable reference object representation allows the spatial semantics to be continuously aligned with the visual scene.

The spatial terms are characterized by the shape of *spatial semantic templates*, which define the connectivity between a particular spatial term and a "retinotopic" space. These connection weights are modulated by the localized activity in the reference field representing the user-defined reference object. Thus, the spatial semantic templates are aligned with the location of the reference object in the image. This shift is accomplished by a convolution of the outcome of the reference field with the spatial semantic template. The particular functions defining "left", "right", "in front", and "behind" here are two-dimensional Gaussians in polar coordinates and are based on a neurally-inspired approach to English spatial semantic representation (O'Keefe, 2003). In Cartesian coordinates, they have a tear-drop shape. The shapes of spatial semantic templates were explicitly designed in this work, but arbitrary shapes can be acquired

7.3 An extension of the spatial language architecture and the need for sequential organization of boosts

in a learning process, by laying down memory traces in the neural fields' dynamics.

The *spatial semantic fields* (see Figure 7.1) are dynamic neural fields with weak lateral interactions. They integrate the spatial semantic user input (aligned with the reference object) with the summed output of the color-space fields. Both these connections are reciprocal. The summed output of the spatial semantic fields serves as input to the color-space fields, enhancing activation in those regions corresponding to the specific spatial term. When active, they also provide input to spatial-term nodes, triggering a spatial-term output ("left", "right", etc.).

A mathematical description of the robotic spatial language architecture can be found in (Lipinski *et al.*, 2009d). Further, I describe the extended spatial language architecture, used to model results of empirical studies and to link more closely to neural mechanisms. The mathematical expressions for that model are presented in Section 7.4, together with the ordinal dynamics that controls the activation boosts in the autonomous version of the model.

7.3 An extension of the spatial language architecture and the need for sequential organization of boosts

Figure 7.2 presents an overview of the neural spatial language architecture, the mathematical description of which is presented in Section 7.4. The previous DFT spatial language architecture is modified to reflect more closely the neurophysiology of the processes involved in spatial language behavior: processing of visual information, spatial representations of referenced and target location, and shift of the reference frame. The camera image (Figure 7.2a) provides visual input to the system. All elements shown in gray in the figure are dynamic neural structures: Gray rectangles (Figure 7.2c, d, e and g) are dynamic neural fields defined over a two-dimensional visual space. Gray circles are discrete nodes representing color terms (Figure 7.2b), spatial relations (Figure 7.2h), or spatial terms (Figure 7.2i) that follow the same dynamic principles as the fields. The transformation field (Figure 7.2f) is a higher-dimensional dynamic neural field that holds a representation of the target object in the reference frame of the reference object. Excitatory interactions between elements are indicated by arrows. These interactions are typically bidirectional in this architecture, shown as double arrows. Diamond-shaped links (Figure 7.2d, e) represent inhibitory projections. The connections between the object-centered field (Figure 7.2g) and the spatial relation nodes (Figure 7.2h) depend on custom semantic weights, shown exemplarily for the ÒaboveÓ relation (Figure 7.2j). The semantic weight patterns describe how well a certain posi-

7. APPLICATION TO CONTROL THE COGNITIVE DYNAMICS WITHIN A DFT ARCHITECTURE FOR SPATIAL LANGUAGE.

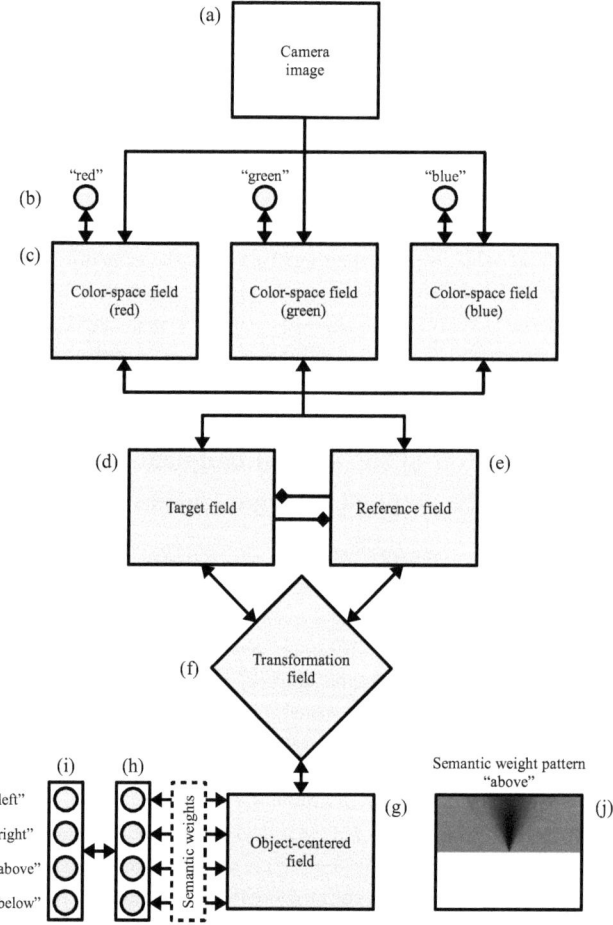

Figure 7.2: Overview of the modified spatial language architecture.

tion in the object-centered field matches the meaning of a spatial term (darker colors

7.3 An extension of the spatial language architecture and the need for sequential organization of boosts

mean higher weights).

Figure 7.3 shows three snapshots of the dynamics of the spatial language architecture, when the robot answers the question "What color has the object to the right of the green object?". In the non-autonomous implementation, to ask this question means, first, to boost the *green* color node and thus increase the resting level of the *green* color-space field. Simultaneously (or after a peak has been built in the color-space field), the *reference* field is boosted, so that a peak can be built there over the location of the green object in the scene (Figure 7.3a). Second, the *right* spatial-term node is boosted together with the reference-centered spatial field, where activation builds-up in the right portion of the neural field (Figure 7.3b). After that, the *target* field can be boosted and a peak builds-up there at the location of the object, whose representation overlaps most strongly with input from the reference-centered field. At last, all color-nodes are boosted to force a decision among them for the color of the object represented in the target field, *blue* here (Figure 7.3c). The robot produces the right answer and the task is complete.

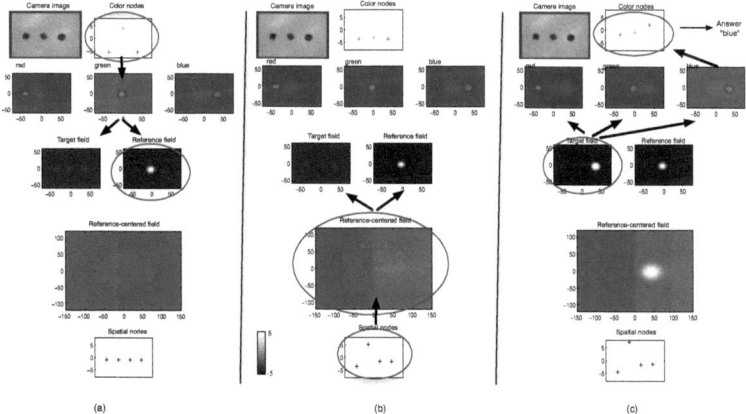

Figure 7.3: Task "What color has the object to the right form the green one". Non-autonomous architecture. Circles around neural fields or nodes denote parts of the architecture that receive a boost in a particular stage of the processing: (a) the reference object is specified; (b) the spatial term is specified; (c) target field is boosted to ultimately obtain the answer, the target's color. Arrows display the flow of the propagating activation.

101

7. APPLICATION TO CONTROL THE COGNITIVE DYNAMICS WITHIN A DFT ARCHITECTURE FOR SPATIAL LANGUAGE.

Figure 7.4 shows another exemplary task, accomplished by the architecture in a non-autonomous fashion. Here, the question "Where is the red object relative to the green object?" is asked by the user. To ask this question, the following sequence of boosts is needed. First, the *red* color node is boosted so that the *red* color-space field receives additional activation. At the same time, the *target* field is boosted and builds a peak of activation over the location of the red object in the scene (Figure 7.4a). Second, the *green* color node is boosted, that provides additional spatially homogeneous input to the *green* color-space field, where a peak is built over the location of the green object. The *reference* field is also boosted now and the induced peak in that field represents the green reference object now (Figure 7.4b). Finally, the *reference-centered* field and *spatial* nodes are boosted and the system produces the correct answer "to the left" (Figure 7.4c).

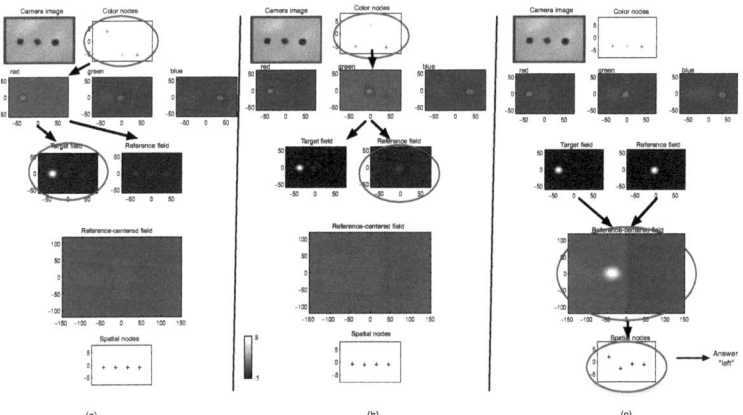

Figure 7.4: Task "Where is the red object relative to the green object?". Non-autonomous architecture. Circles around neural fields or nodes denote parts of the architecture that receive a boost in a particular stage of the processing: (a) the target object is specified; (b) the reference object is specified; (c) the reference-centered field and the spatial nodes are boosted to obtain the answer - a spatial term. Arrows display the flow of activation.

7.4 Mathematical description of the modified architecture with the ordinal component

Here, I briefly summarize the mathematics of the architecture for grounded spatial language.

7.4.1 Color-space fields

Activation of the color-space fields within the spatial language architecture develops according to Eq. 7.1.

$$\tau_v \dot{u}_{v,i}(\mathbf{x},t) = - u_{v,i}(\mathbf{x},t) + h_v + \iint f(u_{v,i}(\mathbf{x}',t))\omega_{2D} d\mathbf{x} + c_{cv} f(d_{c,i}(t)) + c_{tv} f(u_t(\mathbf{x},t))$$
$$+ c_{rv} f(u_r(\mathbf{x},t)) + \sum_{j=0}^{N_{ord}} C_{v,i,j}(t) d_j(t), \quad i = 0..N_{col}, j = 0..N_{ord}$$
(7.1)

Here, $\mathbf{x} = [x,y]$ denotes the two-dimensional visual space, i.e. the horizontal and the vertical dimensions of the image plane. Index i numbers the discrete color categories, represented in this coarse color-space field implementation. $\omega_{2D} = \omega_{2D}(x - x', y - y')$ is a two-dimensional Gaussian kernel with local excitatory part and a global inhibitory part. Parameter c_{cv} controls strength of input from the color-space node, corresponding to the particular color-space field. Parameters c_{tv} and c_{rv} control contributions from the target and the reference dynamic neural fields respectively. The connection weights $C_{v,i,j}(t)$ connect the ordinal nodes to the color-space fields and provide for a boost of the particular field if a node that is connected to it by positive weights is active. These boosts are used to set-up a particular task, specifying the color of the target or the reference object. A particular set of weights across the whole architecture is chosen according to the user input.

7.4.2 Target field

The target field receives input from the visual field when the color named by the user refers to the target object of the spatial language task. Activation of the target field

7. APPLICATION TO CONTROL THE COGNITIVE DYNAMICS WITHIN A DFT ARCHITECTURE FOR SPATIAL LANGUAGE.

follows the dynamics of Eq. 7.2.

$$\tau_t \dot{u}_t(\mathbf{x_2}, t) = -u_t(\mathbf{x_2}, t) + h_t + \iint f(u_t(\mathbf{x_2'}, t))\omega_{2D} d\mathbf{x_2}$$
$$+ c_{vt} \sum_{i=0}^{N_{col}} f(u_{v,i}(\mathbf{x_2}, t)) + c_{mt} \iint f_2(u_m \mathbf{X}, t)) d\mathbf{x_1} + \sum_{j=1}^{N_{ord}} C_{t,j}(t) d_j(t) \quad (7.2)$$

Here, c_{vt} controls the strength of the input from visual color-space fields, summed over colors. c_{mt} is the parameter defining the strength of input from the transformation field, $u_m(\mathbf{X}, t)$. $C_{t,j}$ are the connection weights from the ordinal nodes, as in Eq. 7.1. Other notation is analogous to that of Eq. 7.1.

7.4.3 Reference field

Notation and dynamics of the reference field is analogous to that of the target field, Eq. 7.2:

$$\tau_r \dot{u}_r(\mathbf{x_1}, t) = -u_r(\mathbf{x_1}, t) + h_r + \iint f(u_r(\mathbf{x_1'}, t))\omega_{2D} d\mathbf{x_1'} + c_{vr} \sum_{i=0}^{N_{col}} f(u_{v,i}(\mathbf{x_1}, t))$$
$$+ c_{mr} \iint f_2(u_m(\mathbf{X}, t)) d\mathbf{x_2} + \sum_{j=1}^{N_{ord}} C_{r,j} d_j(t) \quad (7.3)$$

7.4.4 Transformation field

The mechanism of transforming the representation of the target object from the image-based reference frame into a coordinate frame anchored on the reference object is described in (?). The activity in the transformation dynamic neural field in this implementation develops according to Eq. 7.4.

$$\tau_m \dot{u}_m(\mathbf{X}, t) = -u_m(\mathbf{X}, t) + h_m + \iiiint f(u_m(\mathbf{X'}, t))\omega_{4D} d\mathbf{X'}$$
$$+ c_{tm} f(u_t \mathbf{x_2}, t)) + c_{rm} f(u_r(\mathbf{x_1}, t)) + c_{obj,m} f(u_{obj}(\mathbf{x_o}, t)) \quad (7.4)$$

The transformation field is a four-dimensional dynamic neural field that is defined over two spatial dimensions of the reference field, denoted $\mathbf{x_1}$ here and two spatial dimensions of the target field, $\mathbf{x_1}$ ($\mathbf{X} = [\mathbf{x_1}, \mathbf{x_2}] = [x_1, y_1, x_2, y_2]$). Interaction within this field is weak, this field only transfers activity, no detection or selection instabilities happen in this field. It receives input from the target (strength c_{tm}), the reference

7.4 Mathematical description of the modified architecture with the ordinal component

(strength c_{rm}), and object-centered (strength $c_{obj,m}$) neural fields along the corresponding dimensions that is repeated along the other two dimensions for the target and reference fields' inputs and is repeated along the diagonal of the transformation field for the object-center field's input.

7.4.5 Object-centered field

The object-centered field represents the location of the target object relative to the reference object and is connected to the spatial term nodes, that represent these terms relative to the reference, Eq. 7.5.

$$\tau_{obj}\dot{u}_{obj}(\mathbf{x_o}, t) = -u_{obj}((\mathbf{x_o}, t) + h_{obj} + \iint f(u_{obj}(\mathbf{x'_o}, t))\omega_{2D,o}d\mathbf{x'_o}$$
$$+ c_{m,obj}\iint f(u_m(\mathbf{X}, t))d\mathbf{x_o} + c_{sp,obj}\sum_{k=0}^{N_{sp}} f(d_{sp,k}(t))M_{sp,k}(\mathbf{x_o}) \quad (7.5)$$
$$+ \sum_{j=1}^{N_{ord}} C_{obj,j}(t)d_j(t)$$

The peculiarity of this neural field is that it is defined over space in the coordinate frame of the reference object, i.e. $\mathbf{x_o} = [x_o, y_o]$ is a two-dimensional space, on which $x_o = x_2 - x_1, y_o = y_2 - y_1$. The new terms in this equation are the projection (along the diagonal) of the transformation field, $u_m(\mathbf{X}, t)$, with strength $c_{m,obj}$ and input from the spatial-term nodes, $d_{sp}(t)$, with strength $c_{sp,obj}$. This latter input is not homogeneous within the object-centered field, but is shaped by the spatial templates, $M_{sp,k}(\mathbf{x_o})$, that define the spatial distribution that corresponds to the kth spatial term. This distribution can be learned in an interaction with the robot, and were set to tear-drop shapes here (Gaussians in polar coordinates).

7.4.6 Spatial and color nodes

The spatial and color nodes are discrete dynamical node that follow the dynamics of Eq. (7.7).

7. APPLICATION TO CONTROL THE COGNITIVE DYNAMICS WITHIN A DFT ARCHITECTURE FOR SPATIAL LANGUAGE.

$$\tau_c \dot{d}_{c,i}(t) = -d_{c,i}(t) + h_c + c_c^{exc} f(d_{c,i}(t)) + c_{vc} \iint f(u_{v,i}(\mathbf{x},t)) d\mathbf{x} \quad (7.6)$$
$$+ \sum_{j=1}^{N_{ord}} C_{c,i,j} d_j(t) + I_{c,user}(t)$$
$$\tau_{sp} \dot{d}_{sp,k}(t) = -d_{sp,k}(t) + h_{sp} + c_{sp}^{exc} f(d_{sp,k}(t)) + c_{vs} \iint M_{sp}(\mathbf{x_o},t) f(u_{obj,k}\mathbf{x_o},t)) d\mathbf{x_o}$$
$$+ \sum_{j=1}^{N_{ord}} C_{sp,k,j} d_j(t) + I_{sp,user}(t) \quad (7.7)$$

7.4.7 Ordinal dynamics

The ordinal and memory nodes follow the dynamic equations 7.8 and 7.9, which are the same as those presented and analyzed in Chapter 5 of the thesis.

$$\tau \dot{d}_i(t) = -d_i(t) + h_d + c_0 f(d_i(t)) - c_1 \sum_{i' \neq i} f(d_{i'}(t)) + c_2 f(d_{i-1}^m(t))$$
$$-c_3 f(d_i^m(t)) - I_C(t) \quad (7.8)$$
$$\tau \dot{d}_i^m(t) = -d_i^m(t) + h_m + c_4 f(d_i^m(t)) - c_5 \sum_{i' \neq i} f(d_{i'}(t)) + c_6 f(d_i(t)) \quad (7.9)$$

The condition of satisfaction node has the dynamics:

$$\tau_{cos} \dot{u}_{CoS}(t) = -u_{CoS}(t) + h_{CoS} + c_{CoS}^{exc} f(u_{CoS}(t)) + c_{CoS}^r B_r(t) \int f(u_r(\mathbf{x_1},t)) d\mathbf{x_1}$$
$$+ c_{CoS}^t B_t(t) \int f(u_t(\mathbf{x_2},t)) d\mathbf{x_2}$$
$$+ c_{CoS}^{obj} B_{obj}(t) \int f(u_{obj}(\mathbf{x_o},t)) d\mathbf{x_o} \quad (7.10)$$

Here, $B_r(t) = \sum_{j=1}^{N_{ord}} C_{r,j} d_j(t)$, $B_t(t) = \sum_{j=1}^{N_{ord}} C_{t,j} d_j(t)$, and $B_{obj}(t) = \sum_{j=1}^{N_{ord}} C_{obj,j} d_j(t)$ are boosts that are provided to the reference, target, and object-centered field respectively by the ordinal nodes. This is a simplified version of the pre-activation of a condition of satisfaction field by the action field in an embodied implementation of the DFT sequence generation architecture, described in Chapter 5.

7.5 Results: autonomous sequencing of boosts.

Figure 7.5 shows the spatial language architecture with ordinal dynamics wired-up for the task "Where is the red object relative to the green object". The wiring is performed

7.5 Results: autonomous sequencing of boosts.

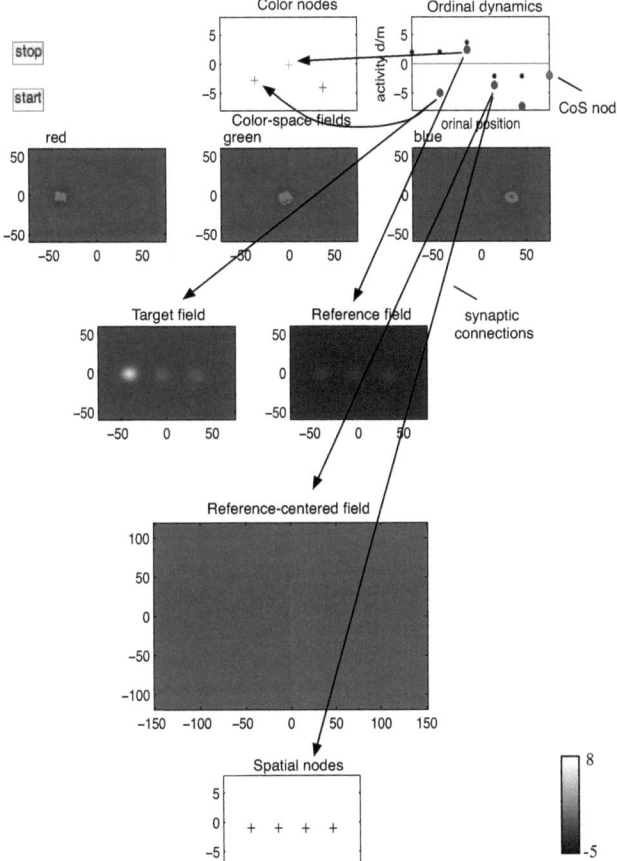

Figure 7.5: Task "Where is the red object relative to the green object?". Autonomous architecture: the synaptic projections from the ordinal dynamics (upper right) specify the sequence of boosts necessary to process the task in the dynamical architecture. See main text for details.

7. APPLICATION TO CONTROL THE COGNITIVE DYNAMICS WITHIN A DFT ARCHITECTURE FOR SPATIAL LANGUAGE.

by adjusting weights of synaptic connections that potentially link every ordinal node to every functional element of the spatial language architecture: color and spatial nodes, target, reference and reference-centered neural fields. The connections can be global, thus directing a homogeneous input to a neural field or a set of neural nodes, or localized, thus setting either a localized input to the neural field, or boosting a single neural node. Here, the connections are set-up separately for each task and the correct set of connection weights is chosen algorithmically depending on the keywords picked-up in the user's question. Emergence of the particular sequences of boost that corresponds to a given spatial language task in a learning process is an interesting direction of future research. In this work, I present how the system functions with autonomous control of dynamic regimes of its subcomponents, but not how the control architecture can be autonomously learned.

To demonstrate that autonomy, I compare the time-courses of the dynamics of color and spatial nodes as well as of the projections on the horizontal axis of the target and reference fields obtained when the architecture was driven by external inputs in a non-autonomous setting and when the ordinal dynamics controlled autonomous switching of boosts. The timing of the boosts can be carefully chosen, so that the spatial language dynamics works fine in the non-autonomous setting (Figure 7.6a). However, if the boosts change too quickly, an error might happen, because a peak in a boosted field didn't have enough time to evolve, or because a new boost was not present long enough to overcome activation caused by the previous boost. Thus, in Figure 7.6b, the system makes an error, mixing up the colors of the reference and target objects, as the boost to the target field was not present long enough for the correct peak over the location of the red object to evolve. That peak was overridden by the input from the green object. The location of the green object was therefore inhibited in the reference field because of the coupling between these fields, and the location of the red object, in its turn, won the competition in the reference field.

In the autonomous setting illustrated in Figure 7.7, such errors are excluded, as the timing of the switches between boosts is controlled by the condition of satisfaction system. The transitions happen only when the desired dynamical states are stabilized. That makes the overall dynamics more robust and the system more flexible, when a perturbation or a competing input slows down the formation of a peak in one of the neural fields in the architecture.

7.5 Results: autonomous sequencing of boosts.

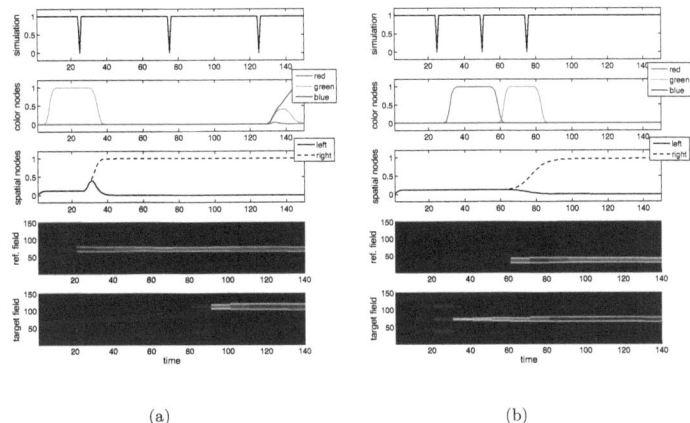

(a) (b)

Figure 7.6: (a) Task "What color has the object to the right of the green object?". (b) Task "Where is the red object relative to the green object?". Boosts are introduced to the system at fixed time intervals.

7. APPLICATION TO CONTROL THE COGNITIVE DYNAMICS WITHIN A DFT ARCHITECTURE FOR SPATIAL LANGUAGE.

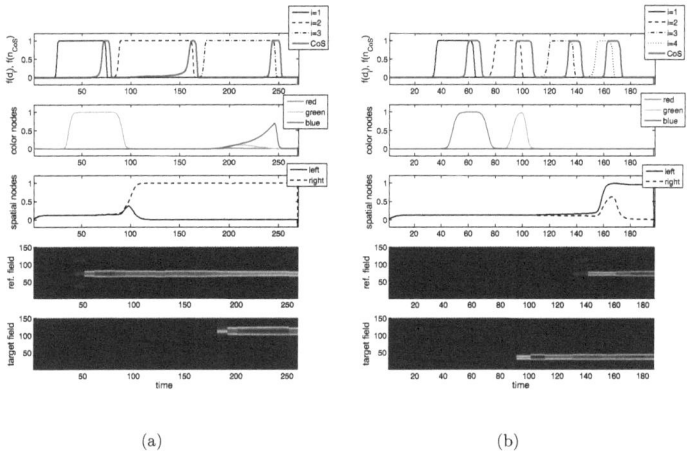

(a)   (b)

Figure 7.7: (a) Task "What color has the object to the right form the green object?". (b) Task "Where is the red object relative to the green object?". Boosts are introduced by the sequence generating system when the conditions of satisfaction for the particular boost is met, i.e. a stable peak has emerged in the boosted neural field.

7.6 Discussion

In this Chapter, I demonstrated how the sequence generation model can be applied to control switching of dynamic regimes in a cognitive architecture. Controlling this switching by a dynamical mechanism makes the overall dynamics more robust, flexible, autonomous, and provides a substrate for learning of the control architecture within dynamic fields framework. This may facilitate understanding of developmental processes underlying emergence of complex cognitive functions and executive control. The properties of the DFT sequence generating architecture were again demonstrated here: stability, flexible timing, gradedness and continuity of representations. Autonomy played a crucial role in enabling autonomous regime changes in the neurally inspired model and the robotic architecture for spatial language behaviors. The autonomous spatial language model will enable comparison of the model's outcome with experimental findings in the temporal domain, as well as unify the robotic implementation with the implementation aiming at accounting for empirical findings.

7. APPLICATION TO CONTROL THE COGNITIVE DYNAMICS WITHIN A DFT ARCHITECTURE FOR SPATIAL LANGUAGE.

8
Discussion

8.1 Addressing the goals set for the thesis

The main goals of my thesis were summarized as follows in the Introduction:

1. Introduce and analyze a model for sequence generation in Dynamic Field Theory (DFT) that solves in a principled fashion the fundamental problem of the stability vs. sequentiality trade-off.

2. Develop and implement an architecture based on the DFT sequence generation model that takes into account constraints of the embodiment on an autonomous robotic agent.

3. Demonstrate in robotic experiments the core properties of the DFT sequence generation model.

4. Apply the DFT model of sequence generation to solve problems of serial order in cognitive tasks.

The main results of the thesis are summarized in the following. The enumeration corresponds to the goals listed above:

1. Analysis of the problem and the model

    - On the basis of an evaluation of the literature on serial order and sequence generation in general and an understanding of the concept and implications of embodiment, formalized in the dynamical systems approach to cognition (Schöner, 2008), in particular, I have proposed a set of constraints for embodied sequence generation. These must be addressed by models of

8. DISCUSSION

sequence generation in order to enable coupling to a real-world behavior through sensors and actuators: autonomy, stability, gradedness, continuous time dynamics, and flexible timing. Of the many models of sequence generation in the literature, some are closely tied to neurophysiology (Aldridge & Berridge, 1998; Carpenter *et al.*, 1999; Procyk *et al.*, 2000; Tanji, 2001) and typically aimed at accounting for particular neural observations within particular experimental paradigms. Other models are constrained by empirical behavioral studies of a particular kind and model serial order errors (Henson, 1998), reaction times (Pfordresher *et al.*, 2007), or memory capacity (Deroost *et al.*, 2006). The result of this approach is a large number of different models, some in conflict with each other, some very general, others more specific. Mathematically, some models use concepts of transient dynamics (Rabinovich *et al.*, 2008), other models introduce semi-stable states (Deco & Rolls, 2005). Neural network models (Botvinick & Plaut, 2004; Elman, 1990) and complex cognitive architectures (Grossberg & Pearson, 2008) were introduced. Notably, none of the models so far has addressed its objectives while at the same time establishing the possibility of embodiment of the resulting architecture, that is the feasibility of coupling to the physical world through a behavior. Robotic sequence generation systems (Alami *et al.*, 1998; Maes, 1989; Payton *et al.*, 1990) naturally offer such coupling, but do not offer means to understand the underlying cognitive, neural, or dynamical mechanisms. The DFT sequence generation model is the first step towards a thorough theoretical framework within which integration of all known constraints on sequence generation and, most importantly, its embodiment can be achieved.

- Using my knowledge of empirical constraints from behavioral and neurophysiological experiments, I have identified a set of necessary dynamic elements that enable reliable switching between two attractor states and their sequencing. These include the representation of the intention of an action ("action", or "motor" field'), the representation of the condition of satisfaction of an action, i.e. its goal or final state ("CoS node or field"), and the representation of the sequential order (in the case when serial order is the mechanism driving sequence generation; other representations of constraints on sequentiality are possible when behavior organization is considered) ("ordinal nodes or stack of ordinal fields"). In my model, these elements are dynamically coupled so that the resulting dynamics guarantees stability of the behavioral states that drive actions of the agent. A transition between the stable attractor states is accomplished through the condition of satisfaction system, which may be coupled to the perceptual system of the agent or to an intrinsic timing mechanism.

- The dynamics of the resulting model was studied in numerical simulations, by observation of time-slices of the phase-portraits of the dynamics (Chapter 5, Section 5.2), and in robotic experiments (throughout the thesis).

2. Development and implementation of a sequence generation architecture

   - the implementation of the model in numerical simulation, and the analysis of its dynamics are discussed in Sections 3.2, 3.3, 5.1, 5.2, 6.2, 6.3;
   - the implementation of the robotic infrastructure necessary for the embodiment of the model is described in Sections 3.4, 5.3, 6.4;
   - the implementation of the model on autonomous robots is described in Chapters 3, 5, 6;
   - the implementation of the model to address questions in cognitive science for turn taking is exemplified in Chapter 4;
   - the implementation of the model to enhance the autonomy of DFT architecture for spatial language is demonstrated in Chapter 7.

3. The core properties of the model – its functioning while coupled to low-level sensory information, resistance to temporal perturbations, autonomy, robustness against noisy sensors and in partially unknown environments – were demonstrated in three sets of robotic experiments, described in Sections 3.5, 5.4, and 6.4.3

4. The application of the model as an element of large-scale cognitive architectures was demonstrated in Chapters 4 and 7

8.2 Overview of the results

The model developed in this doctoral thesis is capable of autonomous sequence generation in a close link to sensory-motor systems. Both cognitive architectures and sensory-motor systems are formulated within the same framework of dynamic neural fields, that rely on continuous representations and continuous-time dynamics on these representations. The attractor states of the dynamics used as representations of actions make it possible to sustain variable and unpredictable durations of actions. The

8. DISCUSSION

mechanism of condition of satisfaction introduced to destabilize the stable states triggers a cascade of instabilities making the transition between attractors a reliable fast transient and decoupling the action system from the controlling dynamics during the transient. This decoupling is guaranteed by a "forgetting" instability, in which activity in an action field that controls the action system ceases. Analysis of the switching dynamics facilitated establishing the dynamic regimes and parameter ranges that enable the sequence generation dynamics.

The thesis is built as follows. After an Introduction into the problem of embodied sequence generation (Chapter 1) and a review of the state of the arts and the background of my study (Chaper 2), I started with introduction of a simple DFT model in Chapter 3, in which a stack of identical neural fields defined over the dimension characterizing actions represented both ordinal information and content of the sequential item. The model was coupled to real sensors of a physical robot and the core properties of the sequencing mechanism were demonstrated in a color-search task performed by an autonomous robot. In Chapter 4, I demonstrated how the simple one-dimensional DFT sequencing model can be applied to model a complex cognitive process of coordination between two partners in a dialogue. The resulting model of turn taking demonstrated flexible and reliable exchange of utterances between two agents coupled through a world model. In Chapter 5, the DFT sequencing mechanism was made discrete, which enabled coupling of ordinal dynamics to action spaces of different behavioral dimensions. The condition of satisfaction system, in its turn, was modeled as a dynamic neural field, specific for the particular action and was coupled to the perceptual system of the agent directly. The modified model was also implemented on a robot and its functionality was demonstrated in the color-search task. Chapter 6 demonstrated that the DFT sequence generation model also works in tasks with a richer behavioral repertoire that includes actions of different modalities. This was demonstrated on a robot in a pick-and-place scenario that involved color search, arm movements, and grasping. The closing Chapter 7 demonstrates how the sequence generation model can enhance autonomy in a dynamic cognitive architecture, exemplified in an application to a DFT spatial language architecture.

The approach pursued in this thesis is challenging and non-standard: it is neither common practice in cognitive science nor in physics to employ robots to study and demonstrate properties of a model. However, when the topic of study is the embodiment of a cognitive architecture, an implementation of the model in a working system becomes a necessary component of the research process. As Richard Feyman once said "If I canÕt build it, I donÕt understand it". A robotic implementation is, on the one hand, a proof of concept, demonstrating that the model indeed can guide autonomous behavior. On the other hand, a robotic implementation is sometimes the only means to reveal conceptual problems of a model. The example of sequence generation is illuminating here, as the problem of variable and unpredictable duration of actions is easily overlooked if the only outcome of the model is numerical data from a disembodied simulation. The noisiness of the sensory input, the fact that input to the neural system is unsegmented, and the problems of categorization and decisions about the presence of a particular object in the visual stream are also easily overlooked in disembodied considerations. The latter problems are particular challenging because they have to be solved reliably and dynamically in real time. Coupling to the world both through perception and through actions closes the processing loop. This means that feedback and recurrence are unavoidable in an embodied system, even if in the information processing model itself only the feedforward flow of information is considered. Robust function is then naturally connected to stability in the dynamical systems sense. For these reasons, I argue that robots are useful tools in the analysis of models for cognitive functions. The dynamical systems approach and DFT offer the means to link cognitive modeling to robotics, enabling the application of robotics as such a tool of cognitive science.

8.3 Outlook

In my thesis, a model of sequence generation is introduced and tested in several robotic implementations and as an element of other dynamic cognitive architectures. Quantitative modeling of experimental findings, both neuronal and behavioral, was not part of the thesis. However, the DFT sequence generation model offers means to address behavioral experimental findings. For instance, the gradedness of dynamic neural fields enables accounting for the effect of similarity known in serial order errors literature. Memory effects (recency and primacy) can be addressed by studying a dynamics of the memory trace that has both build-up and forgetting components. The separation of the ordinal dynamics from the contents of a sequence allows to address serial order errors when noise is introduced to the dynamics of the ordinal nodes, and when failures of the condition of satisfaction signal during learning and production of a sequence are considered. The same mechanisms could account for deteriorating performance with increasing speed of production, whereas effects of chunking in overlearned sequences can be modeled by neuronal learning mechanisms. The mapping of the functional dynamical elements of the architecture to neuronal structures (introduced in Sections 5.5,

8. DISCUSSION

2.1.2, 5.1) makes it possible to align the model more closely with the neurophysiology of sequence generation. Both of these directions of study, neuronal and behavioral, are large-scale projects, which may be carried out in collaboration with experimental scientists in the respective fields.

Behavior organization and goal-directed sequence planning are two other mechanisms that may underly sequence generation in humans. Behavior organization takes into account the constraints on the order of actions that arise from the embodiment of an agent, e.g. the particular configuration of its effectors, from the task setting, or from environmental constraints. Examples of such constraints are the following: one cannot grasp an object before locating it, one cannot lift an object before grasping it, a robot cannot at the same time close and open the gripper. Many of these constraints can be automatically satisfied by a correct choice of representation of the action system. Thus, actions that exclude each other are often connected to the same effector and correspond to different values of the control variable for this effector. Representing such action in the same dynamic neural field with global inhibition excludes their simultaneous activation. To represent other constraints, in particular, task-specific ones, additional neural and dynamical mechanisms are needed. These mechanisms are needed in addition to the elementary mechanism of sequential switching. I therefore excluded these additional mechanisms from consideration in this thesis to make the task tractable. However, one of the possible next steps in the development of the sequence generation framework is to introduce rules of behavior organization into the current DFT sequence generation architecture. This would include the introduction of preconditions of actions along with goal states (or conditions of satisfaction), the introduction of a concept of motivation to signal the need of a particular state and the knowledge that a particular action will lead to this state. In this more general case, not all actions may need an explicit condition of satisfaction (for instance, the action "avoid obstacles"). My argument that one cannot construct sequences of such actions is still true. Such actions can simply run in parallel with other actions for which the conditions of satisfaction are defined and which can be sequentially organized. In a multimodal extension of the DFT sequence generation model, actions can also be executed in parallel if they are coupled to different effectors. In behavior organization one might need to define explicitly whether actions can run in parallel or not, as this might depend on the task. When an "on-going" action prohibits activation of another action, a mechanism must be introduced that defines how one action can "switch off" another one that otherwise stays active forever (as there is no intrinsic condition of satisfaction to switch it off). This will again lead to a generic condition of satisfaction mechanism, something like a "termination condition".

Thus, a system of behavioral organization may be implemented based on the concepts developed in this thesis, including the sequence generating architectures introduced here. Understanding the dynamics of sequential transitions may turn out to be a necessary step towards the development of a fully-fledged control system for an autonomous robot. The problem of how a sequence leading to a particular goal can be elaborated by the agent may then be solved within the behavior organization architecture by propagating a utility function or activation within a motivational system.

Other, more straight-forward extensions of the model include representing subsequences and chunking, coordinating multiple stored sequences, incorporation of an internal timing mechanism in the sequencing architecture, learning of the timing structure of a sequence, autonomous segmentation of action streams during learning, as well as applying the mechanism of sequential switching of dynamic regimes to other cognitive architectures.

8. DISCUSSION

9

Materials & methods

The computational work for the thesis was performed by the author using programming languages Matlab and C++. Implementation of robotic interfaces was accomplished in C++ and was build on top of interfaces available at Institut für Neuroinformatik. Several Open Source libraries were used to preform image processing, visualization, and implement graphical user interfaces. All software developed during work on the thesis can be provided by the author on request. The following table lists parameters of the dynamical architectures, used in the implementation.

9. MATERIALS & METHODS

9.1 Parameters and their values used in implementations the sequence generation model

Name	Value	Name	Value
τ	10.0 [ms]	τ_{nav}	80.0 [ms]
η	20.0 [ms]	τ_{vel}	30.0 [ms]
τ_h	20.0 [ms]	τ_g	20.0 [ms]
h_d	-5.0	λ_{tar}	0.2
h_m	-2.0	λ_{obs}	1.0
h_A	-1.8	β_1	15.0
h_C	-2.0	β_2	30.0
$c_{inh}^{A,C}$	5.0	Ψ_i	[-90, -45, -9, 11, 45, 90] [deg]
β	100.0	v_{tar}	60.0 [pulse/s]
$c_{exc}^{A,C}$	0.9	v_{obs}	30.0 [pulse/s]
$\sigma^{A,C}$	3.0	R	100.0 [mm]
c_{CoS}	15	$\Delta\Psi$	45.0 [deg]
c_{vis}	2.0	h_P	-2.0 / -1.0 (learning)
$c_{vis,C}$	1.0	c_A^P	2.5
c_{ord}	10.0	c_{vis}^P	2.0
c_0	7.2	c_{inh}^P	0.5
c_1	3.6	σ^P	[3.0, 3.0]
c_2	0.9	c_{exc}^P	0.6
c_3	0.8	c_p	1.0
c_4	3.5	c_A	1.5
c_5	2.0	N	180
c_6	2.6	N_{sp}	[160, 120]
γ	10.0	μ	25.0

Table 9.1: Numerical values of parameters used in implementations of the sequence generation model. Units are displayed where applicable.

9.2 Parameters and their values used in the implementation of the spatial language architecture with ordinal dynamics

Name	Value	Name	Value
τ_c	10.0 [ms]	τ_r	80.0 [ms]
τ_{sp}	10.0 [ms]	τ_t	30.0 [ms]
τ_m	10.0 [ms]	τ_{obj}	20.0 [ms]
τ_{CoS}	10.0 [ms]		
h_c	-4.0	h_r	-4.0
h_{sp}	-4.0	h_t	-4.0
h_m	-2.0	h_{obj}	1.0
h_{CoS}	-5.0		
c_c^{exc}	2.4	c_{sp}^{exc}	0.25
c_{CoS}^r	1.0	c_{CoS}^t	0.25
c_{cv}	2.0	c_{vc}	0.01
c_{vt}	6.0	c_{tv}	4.0
c_{vr}	6.0	c_{rv}	4.0
c_{tm}	5.0	c_{mt}	0.175
c_{rm}	5.0	c_{mr}	0.0175
$c_{obj,m}$	1.5	$c_{m,obj}$	0.75
$c_{obj,sp}$	0.002	$c_{sp,obj}$	1.0

Table 9.2: Numerical values of parameters used in the implementation of the spatial language architecture with ordinal dynamics. Units are displayed where applicable.

9. MATERIALS & METHODS

References

AFRAIMOVICH, V., ZHIGULIN, V. & RABINOVICH, M. (2004). On the origin of reproducible sequential activity in neural circuits. *Chaos*, **14**, 1123–1130. 17

ALAMI, R., CHATILA, R., FLEURY, S., GHALLAB, M. & INGRAND, F. (1998). An architecture for autonomy. *International Journal of Robotics Research*, **17**, 315–337. 114

ALDRIDGE, J.W. & BERRIDGE, K.C. (1998). Coding of serial order by neostriatal neurons: A "natural action" approach to movement sequence. *Journal of Neuroscience*, **18**, 2777–2787. 11, 60, 83, 114

AMARI, S. (1977). Dynamics of pattern formation in lateral-inhibition type neural fields. *Biological Cybernetics*, **27**, 77–87. 3, 22, 23

ANDERSON, J., BOTHELL, D., BYRNE, M., DOUGLASS, S., LEBIERE, C. & QIN, Y. (2004). An integrated theory of the mind. *Psychological Review*, **111**, 1036–1060. 3

ARBIB, M.A. (1998). *Schema theory*, 830–834. MIT Press, Cambridge, MA, USA. 3

ARKIN, R. & MACKENZIE, D. (1994). Temporal coordination of perceptual algorithms for mobile robot navigation. *IEEE Transactions on Robotics and Automation*, **10**, 276–286. 21

AVERBECK, B.B. & LEE, D. (2007). Prefrontal Neural Correlates of Memory for Sequences. *J. Neurosci.*, **27**, 2204–2211. 11

BECKER, C., KOPP, S. & WACHSMUTH, I. (2004). Simulating the emotion dynamics of a multimodal conversational agent. In E. André, L. Dybkjaer, W. Minker & P. Heisterkamp, eds., *Affective Dialogue Systems*, vol. 3068 of *Lecture Notes in Computer Science*, 154–165, Springer Berlin / Heidelberg. 45

BEISER, D.G. & HOUK, J.C. (1998). Model of cortical-basal ganglionic processing: encoding the serial order of sensory events. *Journal of Neurophysiology*, **79**, 3168–3188. 14

BICHO, E., MALLET, P. & SCHÖNER, G. (2000). Target representation on an autonomous vehicle with low-level sensors. *The International Journal of Robotics Research*, **19**, 424–447. 39

BOTVINICK, M. & PLAUT, D.C. (2004). Doing without schema hierarchies: A recurrent connectionist approach to normal and impaired routine sequential action. *Psychological Review*, **111**, 395–429. 13, 114

BOTVINICK, M.M. & PLAUT, D.C. (2006). Short-term memory for serial order: a recurrent neural network model. *Psychological Review*, **113**, 201–233. 6, 12, 22, 83

BRADSKI, G., CARPENTER, G. & GROSSBERG, S. (1994). Store working memory networks for storage and recall of arbitrary temporal sequences. *Biological Cybernetics*, **71**, 469–480. 83

BROWN, G.D.A., PREECE, T. & HULME, C. (2000). Oscillator-based memory for serial order. *Psychological Review*, **107**, 127–181. 6, 22

BURGESS, N. & HITCH, G.J. (1999). Memory for serial order: a network model of the phonological loop and its timing. *Psychological Review*, **106**, 551–581. 6, 13, 22

CARPENTER, A.F., GEORGOPOULOS, A.P. & PELLIZZER, G. (1999). Motor cortical encoding of serial order in a context-recall task. *Science (Reports)*, **283**, 1752–1757. 11, 60, 114

CLOWER, W.T. & GARRETT, A.E. (1998). Movement sequence-related activity reflecting numerical order of components in supplementary and presupplementary motor areas. *Journal of Neurophysiology*, **80**, 1562–1566. 60

CONRAD, R. (1965). Order error in immediate recall of sequences. *Journal of Verbal Learning and Verbal Behavior*, **4**, 161 – 169. 10

COOPER, R. & SHALLICE, T. (2000). Contention scheduling and the control of routine activities. *Cognitive Neuropsychology*, **17**, 297–338. 5, 12

COOPER, R. & SHALLICE, T. (2006). Hierarchical schemas and goals in the control of sequential behavior. *Psychological Review*, **113**, 887–916. 3

DEADWYLER, S.A. & HAMPSON, R.E. (1995). Ensemble activity and behavior: What's the code? *Science*, **270**, 1316–1318. 6

DECO, G. & ROLLS, E.T. (2005). Sequential memory: A putative neural and synaptic dynamical mechanism. *Journal of Cognitive Neuroscience*, **17**, 294–307. v, 6, 14, 17, 20, 22, 83, 114

DECO, G., JIRSA, V.K., ROBINSON, P.A., BREAKSPEAR, M. & FRISTON, K. (2008). The dynamic brain: from spiking neurons to neural masses and cortical fields. *PLoS Computational Biology*, **4**, e1000092. 20

DELL, G.S., BURGER, L.K. & SVEC, W.R. (1997a). Language production and serial order: a functional analysis and a model. *Psychological Review*, **104**, 123–147. 10

DELL, G.S., CHANG, F. & GRIFFIN, Z.M. (1997b). Connectionist models of language production: lexical access and grammatical encoding. *Cognitive Science: A Multidisciplinary Journal*, **23**, 517–542. 5, 10, 12, 13

DEROOST, N., KERCKHOF, E., COENE, M., WIJNANTS, G. & SOETENS, E. (2006). Learning sequence movements in a homogeneous sample of patients with parkison's desease. *Neurophychologia*, **44**, 1653–1662. 10, 114

DOMINEY, P., ARBIB, M. & JOSEPH, J.P. (1995). A model of corticostriatal plasticity for learning oculomotor associations and sequences. *Journal of cognitive Neuroscience*, **7**, 311–336. 13

EICHENBAUM, H. (2007). Comparative cognition, hippocampal function, and recollection. *Comparative Cognition and Behavior Reviews*, **2**, 47–66. 12

REFERENCES

ELMAN, J. (1990). Finding structure in time. *Cognitive Science*, **14**, 179–211. 6, 13, 22, 84, 114

ERHORUL, C. & EICHENBAUM, H. (2006). Essential role of the hippocampal formation in rapid learning of higher-order sequential associations. *The Journal of Neuroscience*, **26(15)**, 4111–4117. 10

ERICKSON, R.P. (1974). Parallel "population" neural coding in feature extraction. In F.O. Schmitt & F.G. Worden, eds., *The neurosciences — Third study program*, 155–169. 6

ERLHAGEN, W. & SCHÖNER, G. (2002). Dynamic field theory of movement preparation. *Psychological Review*, **109**, 545–572. 32

ERMENTROUT, B. (1998). Neural networks as spatio-temporal pattern-forming systems. *Reports on Progress in Physics*, **61**, 353–430. 3, 22, 23

FARRELL, S. & LEWANDOWSKY, S. (2002). An endogenous distributed model of ordering in serial recall. *Psychonomic Bulletin and Review*, **9**, 59–79. 5

FAUBEL, C. & SCHÖNER, G. (2008). Learning to recognize objects on the fly: a neurally based dynamic field approach. *Neural Networks*, **21**, 562–576. 27, 83

FUJII, N. & GRAYBIEL, A.M. (2003). Representation of action sequence boundaries by macaque prefrontal cortical neurons. *Science*, **301**, 1246–1249. 11

GEORGOPOULOS, A.P. (1991). Higher order motor control. *Annual Reviews of Neuroscience*, **14**, 361–377. 6

GLASSPOOL, D., SHALLICE, T. & CIPOLOTTI, L. (????).

GLASSPOOL, D.D., SHALLICE, T. & CIPOLOTTI, L. (2004). Neuropsychologically plausible sequence generation in a multi-layer network model of spelling. 12

GLASSPOOL, D.W. (2005). Serial order in behaviour: Evidence from performance slips. *Connectionist Models in Cognitive Psychology*, 241. 10

GLASSPOOL, D.W. & HOUGHTON, G. (2005). Serial order and consonant-vowel structure in a graphemic output buffer model. *Brain and Language*, **94**, 304–330. 5

GNADT, W. & GROSSBERG, S. (2007). SOVEREIGN: An autonomous neural system for incrementally learning planned action sequences to navigate towards a rewarded goal. *Neural Networks*, **21**, 699–758. 14

GROSSBERG, S. (1978). A theory of human memory: self-organization and performance of sensory-motor codes, maps, and plans. In R. Rosen & F. Snell, eds., *Progress in theoretical biology*, vol. 5, 233–374, Academic Press, New York. 5, 12, 14

GROSSBERG, S. & PEARSON, L. (2008). Laminar cortical dynamics of cognitive and motor working memory, sequence learning and performance: Toward a unified theory of how the cerebral cortex works. *Psychological Review*, **115**, 677–732. 14, 114

HARTLEY, T. & HOUGHTON, G. (1996). A linguistically constrained model of short-term memory for nonwords. *Journal of Memory and Language*, **35**, 1–31. 5

HENSON, R.N. (1998). Short-term memory for serial order: The start-end model. *Cognitive Psychology*, 73–137. 3, 5, 10, 12, 13, 114

HENSON, R.N.A. & BURGESS, N. (1997). Representations of serial order. In J.A. Bullinaria, D.W. Glasspool & G. Houghton, eds., *Connectionist Representations*, 283–300, Springer Verlag. 43

HIKOSAKA, O., NAKAHARA, H., RAND, M.K., SAKAI, K., LU, X., NAKAMURA, K., MIYACHI, S. & DOYA, K. (1999). Parallel neural networks for learning sequential procedures. *Trends in Neurosciences*, **22**, 464–471. 12

HIKOSAKA, O., TAKIKAWA, Y. & KAWAGOE, R. (2000). Role of the basal ganglia in the control of purposive saccadic eye movements. *Physiol Rev.*, **80**, 953–78. 11

HOUGHTON, G. (1990). The problem of serial order: a neural network model of sequence learning and recall. In R. Dale, C. Mellish & M. Zock, eds., *Current research in natural language generation*, 287–319, Academic Press Professional, Inc., London. 6, 22

HOUGHTON, G. & HARTLEY, T. (1995). Parallel models of serial behavior: Lashley revisited. *Psyche*, **2**, 2–25. 13

JOHNSON, J.S., SPENCER, J.P. & SCHÖNER, G. (2006). A dynamic neural field theory of multi-item visual working memory and change detection. In *Proceedings of the 28th Annual Conference of the Cognitive Science Society (CogSci 2006)*, 399–404, Vancouver, Canada. 54

JOHNSON, J.S., SPENCER, J.P. & SCHÖNER, G. (2008). Moving to higher ground: The dynamic field theory and the dynamics of visual cognition. *New Ideas in Psychology*, **26**, 227–251. 6, 27, 43

JOHNSON, J.S., SPENCER, J.P., LUCK, S.J. & SCHÖNER, G. (2009). A dynamic neural field model of visual working memory and change detection. *Psychological Science*, **20**, 568–577. 27

JORDAN, M.I. (1997). Chapter 25 serial order: A parallel distributed processing approach. **121**, 471 – 495. 12

KEELE, S.W., IVRY, R., MAYR, U., HAZELTINE, E. & HEUER, H. (2003). Cognitive and neural architecture of sequence representation. *Psychological Review*, **110**, 316–339. 3

KELSO, J.A.S. (1995). *Dynamic Patterns: The Self-Organization of Brain and Behavior*. The MIT Press. 4

KELSO, J.A.S. & SCHÖNER, G. (1987). Toward a physical (synergetic) theory of biological coordination. *Springer Proceedings in Physics*, **19**, 224–237. 4

KOSECKA, J. & BAJCSY, R. (1993). Discrete event systems for autonomous mobile agents. *Robotics and Autonomous Systems*, **12**, 187-198. 21

LANDAU, B. & JACKENDOFF, R. (1993). "what" and "where" in spatial language and spatial cognition. *Behavioral and Brain Sciences*, **16**, 217–265. 96

LASHLEY, K.S. (1951). The problem of serial order in behavior. In L.A. Jeffress, ed., *Cerebral mechanisms in behavior*, Wiley, New York. 9

LEE, C.L. & ESTES, W.K. (1977). Order and position in primary memory for letter strings. *Journal of Verbal Learning and Verbal Behaviot*, **16**, 395–418. 10

REFERENCES

LEVINSON, S. (2003). *Space in language and cognition: Explorations in cognitive diversity.* Cambridge Univ Pr. 95

LI, K.Z.H., LINDENBERGER, U., RÜNGER, D. & FRENSCH, P.A. (2000). The role of inhibition iin the regulation of sequential action. *Psychological Science*, **11**, 343–347. 10

LIPINSKI, J., SPENCER, J.P., SAMUELSON, L.K. & SCHÖNER, G. (2006). Spam-ling: A dynamical model of spatial working memory and spatial language. In *Proceedings of the 28th Annual Conference of the Cognitive Science Society (CogSci 2006)*, 768–773, Vancouver, Canada. 27

LIPINSKI, J., SANDAMIRSKAYA, Y. & SCHÖNER, G. (2009a). Behaviorally flexible spatial communication: Robotic demonstrations of a neurodynamic framework. In B. Mertsching, M. Hund & A. Z., eds., *KI 2009, Lecture Notes in Artificial Intelligence*, vol. 5803, 257–264, Berlin: Springer-Verlag. 7, 96

LIPINSKI, J., SANDAMIRSKAYA, Y. & SCHÖNER, G. (2009b). Flexible spatial language behaviors: Developing a neural dynamic theoretical framework. In A. Howes, D. Peebles & R. Cooper, eds., *9th International Conference on Cognitive Modeling, ICCM 2009. Manchester, UK.*. 96

LIPINSKI, J., SANDAMIRSKAYA, Y. & SCHÖNER, G. (2009c). Flexible spatial language behaviors: Developing a new dynamic theoretical framework. In *31st Annual Conference of the Cognitive Science Society, Amsterdam*. 7

LIPINSKI, J., SANDAMIRSKAYA, Y. & SCHÖNER, G. (2009d). Swing it to the left, swing it to the right: Enacting flexible spatial language using a neurodynamic framework. *Cognitive Neurodynamics, special issue on Language Dynamics*, **3**. 7, 27, 96, 99

LIPINSKI, J., SPENCER, J.P. & SAMUELSON, L. (2009e). Towards the integration of linguistic and non-linguistic spatial cognition: A dynamic field theory approach. In J. Mayor, N. Ruh & K. Plunkett, eds., *Progress in Neural Processing 18: Proceedings of the Eleventh Neural Computation and Psychology Workshop*, World Scientific, Singapore. 27, 96

MAASS, W., NATSCHLÄGER, T. & MARKRAM, H. (2002). Real-time computing without stable states: a new framework for neural computation based on perturbations. *Neural Computation*, **14**, 2531–2560. 14

MAES, P. (1989). How to do the right thing. *Connection Science Journal*, **1**, 291–323. 21, 114

MATARIC, M.J. (2002). Sensory-motor primitives as a basis for imitation: Linking perception to action and biology to robotics. In C.L.N. Kerstin Dautenhahn, ed., *Imitation in animals and artifacts*, 391–422, The MIT Press, Cambridge, MA, USA. 3

MELAMED, O., GERSTNER, W., MAASS, W., TSODYKS, M. & MARKRAM, H. (2004). Coding and learning of behavioral sequences. *TRENDS in Neurosciences*, **27**, 11–14. 14

NAIRNE, J. (1991). Positional uncertainty in long-term memory. *Memory and Cognition*, **19**, 332–340. 10

NOWOTNY, T. & RABINOVICH, M. (2003). Spatial representation of temporal information through spike-timing-dependent plasticity. *Physical Review E*, **68**, 011908. 15, 16

O'KEEFE, J. (2003). *Vector grammar, places, and the functional role of the spatial prepositions in English.* Oxford University Press., Oxford. 98

O'REILY, R.C. (2006). Biologically based computational models of high-level cognition. *Science*, **314**, 91–94. 3

PAGE, M. & NORRIS, D. (1998). The primacy model: A new model of immediate serial recall. *Psychological Review*, **105**, 761–781. 6, 12, 22

PARENT, A. & HAZRATI, L.N. (1995). Functional anatomy of the basal ganglia. i. the cortico-basal ganglia-thalamo-cortical loop. *Brain Research Reviews*, **20**, 91 – 127. 12

PAYTON, D.W., ROSENBLATT, J.K. & KEIRSEY, D.M. (1990). Plan guided reaction. *IEEE Transactions on Systems, Man and Cybernetics*, **20**, 1370–1382. 21, 114

PERONE, S., SPENCER, J.P. & SCHÖNER, G. (2007). A dynamic field theory of visual recognition in infant looking tasks. In *CogSci'2007*, 580–585, Nashville, TN. 27

PFORDRESHER, P., PALMER, C. & JUNGERS, M. (2007). Speed, accuracy, and serial order in sequence production. *Cognitive Science*, **31**, 63–98. 5, 10, 114

POTTHAST, R. & BEIM GRABEN, P. (2009). Dimensional reduction for the inverse problem of neural field theory. *Frontiers in Computational Neuroscience*. 20

PRINZ, J.J. & BARSALOU, L.W. (2000). Steering a course for embodied representation. In E. Dietrich & A. Markman, eds., *Cognitive dynamics: conceptual change in humans and machines*, 71 – 77, MIT Press, Cambridge, MA. 95

PROCYK, E. & JOSEPH, J.P. (2001). Characterization of serial order encoding in the monkey anterior cingulate sulcus. *European Journal of Neuroscience*, **14**, 1041–1046. 83

PROCYK, E., TANAKA, Y.L. & JOSEPH, J.P. (2000). Anterior cingulate activity during routine and non-routine sequential behaviors in macaques. *Nature Neuroscience*, **3**, 502–508. 11, 114

RABINOVICH, M., HUERTA, R. & AFRAIMOVICH, V. (2006). Dynamics of sequential decision making. *Physical Review Letters*, **97**, 188103. v, 14, 15, 18, 83

RABINOVICH, M., HUERTA, R., VARONA, P. & AFRAIMOVICH, V. (2008). Transient cognitive dynamics, metastability, and decision making. *PloS Computat Biol*, **4**. 17, 114

REDGRAVE, P., PRESCOTT, T. & GURNEY, K. (1999). The basal ganglia: a vertebrate solution to the selection problem? *Neuroscience*, **89**, 1009–1023. 83

RIEGLER, A. (2002). When is a cognitive system embodied? *Cognitive Systems Research*, **3**, 339–348. 9

ROSENBLOOM, P.S., LAIRD, J.E., NEWELL, A. & MCCARL, R. (1991). A preliminary analysis of the soar architecture as a basis for general intelligence. *Artificial Intelligence*, **47**, 289 – 325. 3

RYAN, J. (1969). Temporal grouping, rehearsal and short-term memory. *the Quaterly Journal of Experimental Psychology*, **21**, 148–155. 10

REFERENCES

SANDAMIRSKAYA, Y. & SCHÖNER, G. (2006). Dynamic field theory and embodied communication. In I. Wachsmuth & G. Knoblich, eds., *Modeling communication with robots and virtual humans*, Lecture Notes in Artificial Intelligence, Vol. 4930, 260–278, Springer. 7

SANDAMIRSKAYA, Y. & SCHÖNER, G. (2008). Dynamic field theory and embodied communication. In I. Wachsmuth & G. Knoblich, eds., *Modeling Communication for Robots and Virtual Humans*, vol. 4930 of *Springer Lecture Notes in Computer Science/ Lecture Notes in Artificial Intelligence*, 260–278, Springer Verlag. 7, 30

SANDAMIRSKAYA, Y. & SCHÖNER, G. (2009). Memorizing and generating inhomogeneous behavioral sequnces in the dynamic field theory: Concepts and robotic demondtrations. In *13th International Conference on Cognitive and Neural Systems, ICCNS*. 7

SANDAMIRSKAYA, Y. & SCHÖNER, G. (2010a). An embodied account of serial order: How instabilities drive sequence generation. *Neural Networks*, **23**, 1164–1179. 7

SANDAMIRSKAYA, Y. & SCHÖNER, G. (2010b). Serial order in an acting system: a multidimensional dynamic neural fields implementation. In *Development and Learning, 2010. ICDL 2010. 9th IEEE International Conference on*. 7

SANDAMIRSKAYA, Y., LIPINSKI, J., IOSSIFIDIS, I. & SCHÖNER, G. (2010). Natural human-robot interaction through spatial language: a dynamic neural fields approach. In *19th IEEE International Symposium on Robot and Human Interactive Communication, RO-MAN*, 600–607, Viareggio, Italy. 7, 96

SCHÖNER, G. (2002). Timing, clocks, and dynamical systems. *Brain and Cognition*, **48**, 31–51. 54

SCHÖNER, G. (2008). Dynamical systems approaches to cognition. In R. Sun, ed., *Cambridge Handbook of Computational Cognitive Modeling*, 101–126, Cambridge University Press, Cambridge, UK. 2, 3, 4, 22, 113

SCHÖNER, G. & DOSE, M. (1992). A dynamical systems approach to task-level system integration used to plan and control autonomous vehicle motion. *Robotics and Autonomous Systems*, **10**, 253–267. 51

SCHÖNER, G. & KELSO, J.A.S. (1988). Dynamic pattern generation in behavioral and neural systems. *Science*, **239**, 1513–1520. 54

SCHÖNER, G., DOSE, M. & ENGELS, C. (1995). Dynamics of behavior: Theory and applications for autonomous robot architectures. *Robotics and Autonomous Systems*, **16**, 213–245. 3, 39, 74, 91

SCHUTTE, A.R., SPENCER, J.P. & SCHÖNER, G. (2003). Testing the dynamic field theory: Working memory for locations becomes more spatially precise over development. *Child Development*, **74**, 1393–1417. 27

SEARLE, J.R. (1983). *Intentionality — An essay in the philosophy of mind*. Cambridge University Press. 30, 60

SELINGER, P., TSIMRING, L. & RABINOVICH, M. (2003). Dynamics-based sequential memory: Winnerless competition of patterns. *Physical Review E*, **67**, 011905. 14, 17

SHIFFRIN, R. & SCHNEIDER, W. (1977). Controlled and automatic human information processing: II. Perceptual learning, automatic attending and a general theory. *Psychological review*, **84**, 127–190. 4

SHIMA, K. & TANJI, J. (1998). Role for Cingulate Motor Area Cells in Voluntary Movement Selection Based on Reward. *Science*, **282**, 1335–1338. 11

SPENCER, J., THOMAS, M. & & MCCLELLAND, J., eds. (2009). *Toward a Unified Theory of Development: Connectionism and Dynamic Systems Theory Re-Considered*. New York: Oxford University Press. 4

SPENCER, J.P. & SCHÖNER, G. (2003). Bridging the representational gap in the dynamical systems approach to development. *Developmental Science*, **6**, 392–412. 3, 4

SPENCER, J.P. & SCHÖNER, G. (2006). An embodied approach to cognitive systems: A dynamic neural field theory of spatial working memory. In *Proceedings of the 28th Annual Conference of the Cognitive Science Society (CogSci 2006)Society (CogSci 2006)*, 2180–2185, Vancouver, Canada. 27

SPENCER, J.P., SMITH, L.B. & THELEN, E. (2001). Tests of a dynamic systems account of the a-not-b error: The influence of prior experience on the spatial memory abilities of 2-year-olds. *Child Development*, **72**, 1327–1346. 27

STEINHAGE, A. & SCHÖNER, G. (1998). Dynamical systems for the behavioral organization of autonomous robot navigation. In M.G.T. Schenker P S, ed., *Sensor Fusion and Decentralized Control in Robotic Systems: Proceedings of SPIE*, vol. 3523, 169–180, SPIE-publishing. 21

STREEK, J. (1993). Gesture as communication i: Its coordination with gaze and speech. *Communication Monographs*, **60**, 275–299. 54

STRINGER, S., ROLLS, E., TRAPPENBERG, T. & DE ARAUJO, I. (2003). Self-organizing continuous attractor networks and motor function. *Neural Networks*, **16**, 161–182. 15

STRINGER, S.M., ROLLS, E.T. & TRAPPENBERG, T.P. (2004). Self-organising continuous attractor networks with multiple activity packets, and the representation of space. *Neural Networks*, **17**. 15

STRINGER, S.M., ROLLS, E.T. & TAYLOR, P. (2007). Learning movement sequences with a delayed reward signal in a hierarchical model of motor function. *Neural Networks*, **20**, 172–181. 15

TANJI, J. (2001). Sequential organization of multiple movements: Involvement of cortical motor areas. *Annual Reviews of Neuroscience*, **24**, 631–651. 11, 114

THELEN, E. & SMITH, L.B. (1994). *A Dynamic Systems Approach to the Development of Cognition and Action*. The MIT Press, A Bradford Book, Cambridge, Massachusetts. 3

THÓRISSON, K.R. (2002). Natural turn-taking needs no manual: Computational theory and model, from perception to action. In B. Granström, D. House & I. Karlsson, eds., *Multimodality in Language and Speech Systems*, 173–207, Kluwer Academic Publishers, Dordrecht, The Netherlands. 46, 55

TOUSSAINT, M. & GOERICK, C. (2010). A bayesian view on motor control and planning. *From Motor Learning to Interaction Learning in Robots*, 227–252. 3

REFERENCES

TYRRELL, T. (1993). The use of hierarchies for action selection. In J.A. Meyer, H.L. Roitblat & S.W. Wilson, eds., *From Animals to Animats 2*, 138–147, MIT Press, Cambridge MA, USA. 21

WENNEKERS, T. (2002). Dynamic approximation of spatiotemporal receptive fields in nonlinear neural field models. *Neural computation*, **14**, 1801–1825. 23

WENNEKERS, T. (2006). Operational cell assemblies as a paradigm for brain-inspired future computing architectures. *Neural Information Processing-Letters and Reviews*, **10**, 135–145. 20

WENNEKERS, T. & PALM, G. (2007). Modelling generic cognitive functions with operational Hebbian cell assemblies. *Neuronal Network Research Horizons*, 247. 20

WENNEKERS, T. & PALM, G. (2009). Syntactic sequencing in Hebbian cell assemblies. *Cognitive Neurodynamics*, **3**, 429–441. 21

WILSON, H.R. & COWAN, J.D. (1973). A mathematical theory of the functional dynamics of cortical and thalamic nervous tissue. *Kybernetik*, **13**, 55–80. 3, 22

WILSON, M. & WILSON, T.P. (2005). An oscillator model of the timing of turn-taking. *Psychonomic Bulletin and Review*, **12**, 957–968. 46

WOLPERT, D.M. & KAWATO, M. (1998). Multiple paired forward and inverse models for motor control. *Neural Networks*, **11**, 1317–1329. 3

ZIBNER, S.K.U., FAUBEL, C., IOSSIFIDIS, I., SCHÖNER, G. & SPENCER, J.P. (2010). Scenes and tracking with dynamic neural fields: How to update a robotic scene representation. In *Proceedings of the International Conference on Development and Learning (ICDL'10)*. 27

ZIBNER, S.K.U., FAUBEL, C., IOSSIFIDIS, I. & SCHÖNER, G. (accepted). Dynamic neural fields as building blocks for a cortex-inspired architecture of robotic scene representation. *Autonomous Mental Development, IEEE Transactions on*. 26

Die VDM Verlagsservicegesellschaft sucht für wissenschaftliche Verlage abgeschlossene und herausragende

Dissertationen, Habilitationen, Diplomarbeiten, Master Theses, Magisterarbeiten usw.

für die kostenlose Publikation als Fachbuch.

Sie verfügen über eine Arbeit, die hohen inhaltlichen und formalen Ansprüchen genügt, und haben Interesse an einer honorarvergüteten Publikation?

Dann senden Sie bitte erste Informationen über sich und Ihre Arbeit per Email an *info@vdm-vsg.de*.

Sie erhalten kurzfristig unser Feedback!

VDM Verlagsservicegesellschaft mbH
Dudweiler Landstr. 99   Telefon +49 681 3720 174
D - 66123 Saarbrücken   Fax   +49 681 3720 1749
www.vdm-vsg.de

Die VDM Verlagsservicegesellschaft mbH vertritt

Printed by Books on Demand GmbH, Norderstedt / Germany